SUCCESSFUL SMALL-SCALE FARMING

Successful
Small-Scale Farming

An Organic Approach

Karl Schwenke

A Down-to-Earth Book
from Storey Publishing

Storey Communications, Inc.
Pownal, Vermont 05261

The information in this book is true and complete to the best of our knowledge. All recommendations are made without guarantee on the part of the author or Storey Communications, Inc. The author and publisher disclaim any liability incurred with the use of this information.

Cover design by Carol Jessop
Text design by Wanda Harper
Cover photograph by Richard W. Brown
Illustrations by Elayne Sears and the author

Printed in the United States by Capital City Press

Second Edition
Seventh printing, April 1994

Library of Congress Cataloging-in-Publication Data

Schwenke, Karl.
    Successful small-scale farming : an organic approach / Karl
Schwenke. — 2nd ed.
      p.   cm.
    Includes bibliographical references and index.
    ISBN 0-88266-643-6. — ISBN 0-88266-642-8 (pbk.)
    1. Agriculture—Handbooks, manuals, etc. 2. Organic farming—
Handbooks, manuals, etc. 3. Agriculture—United States—
Handbooks, manuals, etc. 4. Organic farming—United States—
Handbooks, manuals, etc. 5. Farms, Small—United States—
Handbooks, manuals, etc. I. Title.
S501.2.S33 1991
630' .973—dc20   90–50417
                            CIP

To Walter R. Hard, Jr., whose patience and good friendship
are appreciated; and to my stepfather, Charles McCash,
whose dedication to agriculture prompted me
to look deeper.

CONTENTS

Preface

Times change. When I first wrote *Successful Small-Scale Farming*, eleven years ago, an "organic" farmer was synonymous with a "lonely hippie troublemaker." Today he is classed somewhere between a high-priced elitist and an opportunistic liar. Whatever the appellation, and I am not particularly fond of these alternatives, the organic farmer is no longer lonely. In fact, the "natural food" bandwagon is now so crowded with claimants it threatens to overturn from all the fat corporate cats that have trampled their way on board.

This revision is *still* designed to help you to grow good, healthy crops. As in the first edition, I have eschewed a chapter on raising livestock because livestock are, at best, an indirect cash crop — that is, one must first grow plants to feed the animals. Caring for soils and growing plants is fundamental to farming, whether you choose to pyramid your farming enterprise or not. My earlier decision to deemphasize livestock has been modestly vindicated in that consumers are buying less red meat, and all indications are that this trend will continue.

My crystal ball got a little cloudier when I predicted that local consumers would support growing organic produce — they do not, at least not in the quantity to sustain significant growth in small-scale agriculture. Don't get me wrong: there are a rare few small-scale farmers who have managed to make it locally. To do it, they have carefully assessed the needs and extent of the local market, factored in their farm location, been sufficiently capitalized, and chosen their crops accordingly. The secret of their success has invariably been to correctly assess their "niche" in the marketing of their product, *and then to fill that niche completely!*

But relying only on local markets is too risky for most would-be small-scale farmers today. They need to go beyond the local market. Ironically, while marketing farm product is the nemesis of the small-scale farmer, it is also his salvation. Innovative marketing ploys are the answer, and I discuss them in the chapters having to do with cash crops.

When young, idealistic, would-be small-scale farmers ask me today what their chances are, I tell them that times are tougher, but that they can still "make it" if they follow the basic principles laid down here. My advice is as old-fashioned as the plow: come to the land prepared to live simply, work hard, be creative about your marketing, and your small-scale venture will thrive.

Preface to the First Edition

This is a book about small-scale farming. I wrote it specifically for the growing number of *new* farmers (a social category unheard of in this country since the mid-nineteenth century) who are migrating to farm country. It is intended to serve as a practical resource for the beginning cash-crop grower, and it emphasizes the care of soil and the growing of plants.

I have deliberately bypassed a chapter on raising livestock because livestock is, at best, an indirect cash crop — the farmer must first grow plants to feed his animals — and I felt that a chapter would only allow a cursory outline of a subject that warrants far more attention.

Farming is a curious kind of occupation. It is a hodgepodge of knowledge blended with a plethora of skills — all in all, a mish-mash that defies structured analysis. Despite our inability to define farming adequately, we all know what we mean when we say *farmer*. He (somehow we always assume the farmer to be a male) is the avuncular figure who stands, pitchfork at the ready, in the midst of a wheat field. Those running for political office are fond of calling him "the backbone of the nation," or "the salt of the earth." Poets rhapsodize about him as "the last lingering reminder of our links with the soil, the eternal ploughman."

Writing this book has burned these stereotyped images out of my mind. Today I see the farmer as a newcomer, a man (or woman) of urban background who wants to grow food for other people, for reasons that he has not yet been able to sort out. Inexperienced, but willing, he waits for winter to end, the ground to thaw, the chance to begin anew.

It is to this farmer that this book is addressed.

Introduction

The last three decades of American agriculture have been disastrous ones. Lured by cheap, government-subsidized energy sources, we forsook tried-and-true agricultural practices and are now reaping the consequences. Now our soils are becoming dangerously thin, our skilled farm populations continue to move to the cities, and we mass-produce shoddy farm products at exorbitant prices. Once-independent farmers now vie with multinational corporations for room at the public money trough, and once-dedicated husbandrymen now wear the mantle of profiteering agribusinessmen.

It took the energy crisis of the early seventies to make us reexamine this trend (sanity would force us to regard it merely as a trend). Sober evaluation of energy consumption vs. energy production has clearly proven that large, megamechanical farms are hopelessly inefficient, if not downright harmful. The breed of agribusinessmen that this nadir spawned has treated farm soils as an "economic given" — that is, as an exhaustible resource that was theirs by right of investment capital to use up in their quest for more profits.

Any old-time small family farmer can tell you how bankrupt and shortsighted such reasoning is. Small farmers have traditionally valued self-reliance and sound husbandry. This respect was born of the fact that their small plot of land was both the basis of their livelihood, and a valued legacy that they intended to pass on. Historically the small family farm was a labor-intensive, virtually self-contained economic unit, and therein lay its strength. It was an economic unit that in recent analysis proved less energy-consumptive, and more conservative of valuable topsoils than its modern-day successor.

The problems brought about by agribusiness are quite clear, but the solutions are not. I do not believe that we can simply turn the farming clock back to the 1940s. Nor do I think that we can, as Jefferson would have willed it, become a nation made up entirely of small family farmers. But if we are to save our soil and provide for a continuing crop production we can (and must) reduce the scale of our farms to some intermediate level. We must diversify our crops, and undertake immediate and drastic soil conservation methods.

The present practice of distributing California lettuce in Maine and Maine potatoes in California is both wasteful of our nation's energy and ruinous of sound farm practices. Made possible by tax subsidization of the trucking industry, nationwide distribution of perishable farm products has effectively deterred local truck farming, and put consumers at the mercy of middleman speculators. These expensive and wasteful practices cannot continue. Locally produced farm crops will inevitably reassert themselves on the market, and the change will reestablish the need for the small-scale farm.

Reducing scale means more farmers, and because of the generations-old migration from the small farms to the cities, we will need to educate (or reeducate) several new generations of small-scale, intermediate-technology farmers. At this writing we have witnessed a surge of migrants from the cities to the rural areas, and some of these newcomers have tried their hand at small-scale farming. A few have succeeded, but more have failed. Operating mostly on abandoned "sidehill" farms (an economic reality), they attempted to wrest a modest living from the soil. To grow a decent cash crop they have had to contend with thin soils, a paucity of fundamental farming knowledge, and a hostile reception from entrenched agricultural interests — both private and public. Information was hard to come by. Those who still knew how to make a small farm work were either dead or had retired or moved away. Agribusinessmen were no help, because they had become specialists and had lost their ties to the soil.

Availability of fundamental and basic farm knowledge, plus a creative "niche" marketing strategy, more often than not made the difference between success and failure of these new small-scale farm ventures. The overall spectrum of knowledge that the small-scale farmer must acquire is awesome. He must be the proverbial "jack-of-all-trades," and he must be the master of many of them. Aside from a fundamental acquaintance with farm machinery and its specific function, he must have a considerable fund of knowledge in the "-ologies" (biology, entomology, meteorology, pomology, plant pathology, etc.). He needs a working acquaintance with silviculture, veterinary medicine, and basic economics, and the day-to-day requirements of working the soil require him to have a sound background in agronomy. His work also encompasses elements of accounting, surveying, and engineering, and he cannot avoid becoming a passable rough carpenter and a journeyman mechanic.

The latter is particularly appropriate for today's small-scale farmer, because used, small-acreage equipment is at the base of his operation. This equipment is often twenty or more years old and in frequent need of repair. Parts for these machines are sometimes impossible to come by, hence the farmer must turn inventor, designer, and machinist to create the needed parts in his wood or metal shop.

But even as old, used equipment presents problems, it is the present-day small-scale farmer's economic salvation. Agribusiness's emphasis on ever larger, more specialized equipment has flooded today's market with an abundance of machinery designed for use on small acreages. It is an additional bonus that most of this equipment was built before the farm-machinery industry succumbed to the doctrine of planned obsolescence.

While tools play an important role in the small-scale farmer's life, the crops he grows and the health of his soils remain his major concerns. Diversification and long-term rotation of crops are old-time techniques that have a place in the new small-scale farmer's lifestyle and livelihood. The old adage that one shouldn't put all one's eggs in a single basket has special meaning for today's small farmer, because the odds are that he is working on an economic margin that will not sustain a total cash-crop loss. Drought, insect infestations, or crop diseases could, if the fields were devoted to a single crop, wipe out a whole year's work, and thus endanger the whole enterprise.

Handling the problems of stirring the soil, planting, cultivating, harvesting, processing, and storing several different kinds of small-acreage cash crops (diversification) could be a ruinously expensive proposition if one allowed himself to be seduced by specialist equipment advertisements. The alternative is to revert to "antiquated" methods and equipment that substituted domestic sweat for imported energy. This is precisely what most successful new small-scale farmers are doing. In the case of equipment, it is often a question of balancing the versatility of a piece of equipment against the value of the proposed cash crop. But, in any case, good basic tools are an indispensable necessity for the new small-scale farm.

Fortunately, diversification and long-term rotation are practical farm practices for today's small farm. Wholesale large-scale agriculture as it is practiced today dominates the overall consumer market and the larger wholesale outlets. The smaller, more discriminating markets are where today's small-scale farmer sells most of his crops. Since his markets (generally of the direct farmer-to-consumer type) are small, his crop production is geared to match. Although the needed quantity of a specific crop is less, his quality should be higher. Given healthy soils and good basic tools, there is no excuse for the new farmer turning out an inferior product.

But healthy soils are not acquired overnight, and new farmers soon find that soil improvement is a long-term project. A green manure crop, for example, that is grown one year does not yield its benefits in humus and nutrients until the following year (or even later), and mistakes and experiments are often a year in the correcting and evaluating. Crop rotations to bring soils to a fertile state, or to maintain that fertility, are essential, but because they are expensively time-consuming, they must be figured into the small farm's eventual profit or loss.

To the shame of American agriculture, sound conservation practices have become "archaic" in the last twenty years. Indicative of this trend is the U.S. Soil Conservation Service (SCS), which now works as the poor handmaiden of agribusiness. This is not to deny that the SCS once had considerable initial conservation influence. It did, but that influence has waned. Formal cooperator agreements were once the rule in farm country, but they are now the exception. Mass-production agriculture now bypasses them as "profit-inhibitive," "outmoded," and "costly."

But the SCS can still occasionally help the small-scale farmer. Particularly useful are SCS recommendations and help in surveying terraces and ponds, but the small farmer should also avail himself of the agency's advice and planning concerning his soils assessments, crop seed selection, and his contouring and irrigation problems. A free, individualized "Soil and Water Conservation Plan," complete with aerial photographs, soils maps, and soil survey interpretations, still is available from the SCS, and it should be one of the first acquisitions of the new farmer.

Whether SCS-drawn or individually conceived, no farm watershed control plans are complete without consideration of the farm woodlot. Many states offer forest management help with specialists who, like SCS agents, assist new farmers with long-range planning. Unfortunately, some small farms no longer have a woodlot. Recent owners (most often of the absentee variety) have had the land cleared to milk it of immediate profits. The effects of this shortsighted and greedy policy have been sudden and disastrous. Without the forest or windbreak cover, the soils have eroded at an accelerated rate. An astounding 24 billion tons of topsoil are lost worldwide each year to deforestation, and a sixth of that — 4 billion tons — is lost right here in the United States. Here it is washed to the sea, or lodged as dam-filling siltation in reservoirs (taxpayer costs for the latter are estimated to amount to over 50 million dollars each year). Conservative estimates put the cost of soil erosion in this country at well over 50 million barrels of oil each year.

Because of its importance, a woodlot may logically be a prerequisite to the newcomer's purchase of a small farm. Besides holding the soils and providing a welcome windbreak, the woodlot is one of the cheapest and surest energy alternatives for the small-scale farm. Here the farmer exercises the same kind of control that he does with his crops. If his woodlot is cared for with regard for its growth potentials, it can provide a lifetime's worth of heating fuel and farm building and fencing materials.

Seen on an international scale, deforestation is one of the greatest single environmental problems confronting mankind. Consequent wind and water erosion, especially when combined with drought, have accounted for the fall of entire civilizations. Equally dangerous is the growing threat of air pollution in the earth's atmosphere, a condition intensified by a worldwide diminution of forest cover. From these perspectives, it is clear that care/planting/replanting of the small-scale farm woodlot should be at the top of the farmer's list of priorities.

Unfortunately, our government does not agree. Subsidization programs (free trees) that once supported this kind of ecologically responsible thinking is history. One may, in fact, broaden the accusation to say that government help and support for the small farmer have fallen to an all-time low. Homesteading also belongs to history, and subsidies, price-support programs, and other give-away gimmicks belong to mass-production agriculture. If small-scale farming continues to grow in this country, it will owe its success to the native grit of the small farmer, not to the politician.

For many, getting on the land is a bitter struggle, and once there they are discouraged to find that the struggle continues, just to remain. It is a distasteful surprise for many to discover that they did not leave politics behind them when they moved to the farm. Repressive local tax laws soon stir them to support taxation laws based on agricultural productivity rather than developmental potential. As they gain experience they become increasingly aware of the unhealthy infringement into agriculture of faceless conglomerates, and this awareness pushes them into political organizations which oppose tax-loss farming, support enforcement of the farm size limitations implicit in the Reclamation Act of 1902, and support co-ops and their right to bargain collectively with food distributors.

Unlike his homesteading, counterculture neighbor who "works out" for most of his income, the small-scale farmer attempts to wrest his living from his own land, and this makes him sensitive to the effects of government regulation of his markets and the prices being paid for crops of differing quality. He soon learns that most state agencies have abandoned their development function for a "safer" regulatory role, and are, therefore, virtually useless to him. A reading of the daily press tells him that conglomerates today control the prices for at least one-third of all agricultural products, and that this control is being subsidized and encouraged by favorable tax laws.

Coping on a daily basis with infertile soils, pests, and plant diseases, he is made painfully aware of the growing influence of the chemical industry in farming. If he is an organic farmer, he is appalled at the six-fold increase in the amount of synthetic fertilizer used since 1945 (synthetic fertilizers are now an 8 billion dollar industry), and the seventeen-fold increase in the use of pesticides for the same period. When he seeks information from publicly supported land grant college research sources on ways to substitute for these chemicals, he often finds them unresponsive, uncaring, and irretrievably entangled with agribusiness.

To compound the irony, the organic farmer finds that, while public support ostensibly has increased

for chemical-free food, the average consumer does not back his rhetoric with his buck. When, for example, offered a choice between fruits that are slightly blemished, and organically grown, and those that are cosmetically clean, and chemically raised, Joe Blow Consumer will usually choose the latter. Organic farmers, therefore, have to be especially careful in selecting the crop(s) they grow and, even more importantly, they must be equally selective and creative about their marketing scenarios.

For all the above reasons, the small-scale farmer is now forced to become a manipulative marketer and a political animal. Given these facts, one cannot help but be amazed at the renaissance of interest in small-scale farming, but it is happening. A recent Gallup Poll revealed that 36 percent of those city people asked said that they would move from the cities if they had the chance. Their reasons for this disenchantment were many, but foremost among them was fear — fear of pollution, unhealthy foods, and urban crime. Fear has a way of reducing man's concerns to basics, and nothing is more basic than farming.

Whatever the reasons, demographic analysts are recording a migrational return to the land, and it is clear that official predictions of the death of small-scale farming were premature.

CHAPTER 1

Soils

The basic unit the farmer works with is his soil. He must, therefore, have a working knowledge of its characteristics and needs. This does not mean that he must possess an advanced degree in agronomy, but he does need to be able to identify his soils and assess their state of health in order to plan intelligently the use of his land.

TOPSOIL

When we speak of soil as a growing medium, we are talking about the uppermost, minute layer of the earth's crust. Farmers call it topsoil, and soil scientists call it the "A horizon." There is, of course, a layer of "subsoil" (the "B horizon") below this that lends long-term mineral nutrients to plants growing in the topsoil, but it is the uppermost layer where most of the action is taking place, and it is this layer which should properly occupy most of the farmer's thought, time, and effort.

Topsoil is an exhaustible resource. It can be abused to the point of virtual sterility, or it can be physically lost — worn away by erosive agents such as wind and water. There is an end to the amount and quality of topsoil available to us, and with our present practices we are using it up at an alarming rate. In the Midwest, for example, pioneers broke the virgin sod of a topsoil layer that averaged 3 feet in depth. Today in many areas of this same region, the farmer's plow turns up subsoil at a meager 6 inches. To correct this erosive trend we need to reexamine the basic nature of soil and our attitudes toward it.

The farmer must begin to look at his topsoil as a fragile, *living* medium for growing his crops. Healthy soil is not inert, it is pulsatingly alive. Bacteria and fungi (the lowest forms of plant life) teem in good topsoil as they convert organic waste materials into usable nutrients for other plants. Life and death cycles in the soil's dense microorganism population (billions to the gram) are essential ingredients to all life on

5

the face of the planet. Without these minute forms of life, the soil is of no use to the farmer — or to anyone else.

But soil organisms are only one part of the delicate balance that the farmer must maintain. The basic framework of all soil is mineralized particles of rock. These are mixed in varying quantities with the other essential ingredients of healthy soil: water, air, and decaying organic matter. The quantities of water, air, and decaying organic matter can be controlled to a certain extent by the farmer, but for the most part he has to make do with his basic soil *texture*. Texture here means the relative coarseness or fineness of the soil as determined by the amounts and sizes of the rock particles in a given sample.

Relative Sizes of Soil Grains

Light Soils	Heavy Soils	
medium sand	very fine sand	fine silt
fine sand	coarse silt	clay

SOIL TEXTURES

Agronomists and soil scientists have developed an elaborate lexicon to describe soil textures, but the farmer, with typical directness, has reduced this to three categories: sandy soil, silty soil, and clayey soil. Mixtures of the three are called loamy soils. Most farmers go by fingertip feel in determining what kinds of soils they are dealing with. They know that sandy soils have coarser, grittier individual particles that do not adhere when wet. Clay, on the other hand, is sticky when wet, and has the finest individual soil particles. Silt has a feel somewhere between the two. It adheres when wet, but is not sticky. Most farmers call it a "velvety" feel.

The size of individual soil particles is important, because their combined surface area affects water retention and circulation, soil temperatures, mineral solubility, growth of microorganisms, and overall fertility. Smaller particles present more surface for water to cling to by surface tension, and more surface area on which minerals can be dissolved. This is a pertinent fact when one realizes that a mere cubic foot of coarse soil presents an astounding 37,700 square feet of surface area.

WATER AND SOIL

Water-holding capacities for various soil textures vary widely. One hundred pounds of clay, for example, can hold 40 pounds of water, while an equal weight of sand may contain only 5 pounds. Water circulates less readily, however, in clay than in sand, because the spaces between the particles are smaller and the clay particles tend to adhere to one another in clumps ("peds") that inhibit circulation.

For the farmer, the practical effects of his soil's water-holding capacity are considerable. His agricultural practices and his long-range planning must take this factor into account. Clay soils, because of water retentiveness, are difficult to work in the early spring. When plowed or stirred too early they mold and then dry into unworkable clods. In areas where the growing season is short, this is an all-important consideration, and the farmer is justified in using all the drainage tricks he can dig out of his bag.

Sandy soils also present practical problems to the farmer, particularly in regions that have little rainfall. Here the problem is to keep what water is available (through rainfall or irrigation) on the land. Terracing and contour cropping are the usual mechanical practices that farmers use to compensate for their soil's porosity.

As a rule of thumb, the farmer would say that finer soils were more fertile, but that they are harder to manage. Holding more water, the smaller-particled soils, like clay, are "heavier" for him to work. This takes a toll of his equipment and uses up valuable work time in the spring. Then he finds that, once the seedbed is prepared, the wetter (finer) soils warm up less readily than the coarser ones. If the growing season is short, these slower-warming soils can be a problem.

There are, of course, exceptions to this rule. Very black soil, though fine and possessed of a high water capacity, can absorb more heat than lighter-colored ones. Topography (slope exposure, ground sags, etc.) and organic material also can play a part in heating soils to the necessary minimum temperatures (45° to 50°F. — 7° to 10°C.) required to begin the growth of a cash crop.

But the chief factor in a soil's temperature still is its water-retention capacity. It takes, for example, about twenty units of heat to raise the temperature of 100 pounds of dry soil one degree. By contrast, it takes five times that amount of heat to raise the same weight of water one degree. And to evaporate those 100 pounds of water, it requires a striking 96,600 heat units.

ORGANIC MATTER

To this point we have discussed only the mineralized rock particles that make up the skeleton of most of our topsoils. Organic matter — mostly of plant origin — plays an equally important role. In fact, in two special classes of soil (peat and muck), decaying organic material is the primary ingredient. Peat has more than 65 percent organic material, and muck from 25 to 65 percent. The balance of both classes is fine sand, silt, and clay.

The presence of organic material in a given sample of soil is critical to the farmer for three reasons:

1. It is indicative of the soil's *tilth* (looseness, mellowness).
2. It is a visible measure of essential soil elements.
3. It assures the farmer that his soil will have a water-holding capacity.

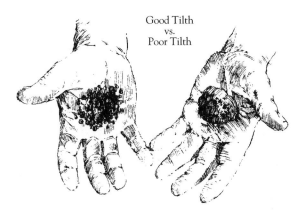

Good Tilth
vs.
Poor Tilth

Soil particles that are covered with black decaying organic matter promise a high nitrogen content, and are a good aerated medium for crop growing. Absence of organic material indicates that crops are likely to suffer poor root growth (lack of aeration), stunted growth due to fewer necessary soil organisms such as bacteria, molds, and fungi, and diminished yield because of a smaller supply of plant food elements.

SOIL ELEMENTS

Soil scientists are still learning about essential plant food elements. These include carbon, hydrogen, oxygen, nitrogen, phosphorus, potassium, sulfur, calcium, magnesium, iron, manganese, zinc, copper, boron, molybdenum, and chlorine. Carbon is the most common constituent of all life, and, logically, it is found most often in soil in the form of decaying organic material.

Farmers are most concerned with the "fertilizing elements," nitrogen, phosphorus, and potassium, simply because these three are most commonly deficient in soil. It is of interest that the total amount of these elements found in virgin soil seldom exceeds 5 percent. This translates out to roughly 3,000 pounds of nitrogen, 4,000 pounds of phosphoric acid, and 16,000 pounds of potash (phosphorus and potassium never occur alone in a natural state) in an acre-sized, 7-inch-thick layer of rich farmland.

Crops taken from farmland diminish the field's resident supply of these elements in measurable quantities. For example, a meager 1-acre crop of 14 bushels of wheat (plus straw) would remove about 14½ pounds of nitrogen, 10½ pounds of phosphoric acid, and 14 pounds of potash. The farmer who is simply trying to replace the nutrients that he uses can readily calculate the loss of these elements for the kinds and quantities of crops he raises by using the table on page 102 in the Appendix. Using a little basic math, anyone can see that the acre cited above has enough nitrogen for 200 such crops, phosphoric acid for 400, and potash for 1,000 — all without adding an ounce of fertilizer.

FERTILE SOILS

Unfortunately, making soil fertile isn't simply a matter of just responding to a chemical analysis. If the needed elements are there but not "available" for use by plants, the farmer is banging his head against a stone wall. What makes these elements available to growing plants is:

1. A suitable moisture supply
2. Good tilth
3. Lots of bacteria and fungi
4. Adequate aeration, and
5. A proper pH for the type of crop grown.

A soil's pH is its relative acidity-alkalinity, measured on a logarithmic scale registering 0 to 14. Thus 7 is considered neutral, above 7 is alkaline ("sweet"), and below 7 is acid ("sour"). To test his soils for sweetness or sourness, the farmer can use a commercially available soil test kit, litmus paper, or the old-fashioned (rough) test of adding a few drops of ammonia to a soil sample that has been stirred into a glass of water. This last test will show whether the sample is sweet or sour, but will not give the degree of sweetness or sourness. Litmus paper will react to the soil sample with colors ranging from green (acid) to blue (neutral) to red (alkaline). Comparison color charts distributed with the soil kits or litmus paper then give the tester accurate readings in the range of pH 4.5 to pH 7.5.

The pH scale deserves a little more examination, because there is a danger that new farmers might regard the scale's range as linear — it is not. Each consecutive number represents *ten times* the value of the preceding number as the numbers decrease. Thus pH 6 is ten times as acid as pH 7, pH 5 one hundred times as acid as pH 7, and pH 4 *one thousand times* as acid as pH 7. Why soil scientists came up with this cumbersome scale defies explanation.

A sour soil is one with an overabundance of hydrogen (acid). Hydrogen, a natural by-product of organic decay, is percolated down through the soil by rainwater, where it displaces plant nutrients (alkaline) on the surfaces of the soil particles. The nutrients continue the downward percolation to the hardpan layer where they are lost by drainage. This process is known as *leaching*.

To rid the soil of this acid condition, farmers spread calcium (in the form of crushed limestone) on the soil. In the past, calcium was regarded solely as a fertilizer, and it is only in relatively recent times that we have come to know that it is both a plant nutrient and a soil conditioner. We found that limestone is mostly calcium carbonate, a combination which water will not dissolve but which a mild acid (hydrogen) will. As the hydrogen dissolves, it combines with the lime and becomes calcium bicarbonate and is thus rendered water-soluble. In suspension in the soil water, it then percolates down and drains away.

Since most helpful soil bacteria, particularly those that release nitrogen by decomposing organic matter, cannot tolerate an acid environment, the application of lime could be considered an "indirect fertilizer." Lime also helps these bacteria by making molybdenum that is resident in the soil available to them. The bacteria need this minor element to fix nitrogen properly. Lastly, there is some scientific thought (disputed) that the application of lime releases the needed fertilizer, potash.

But it is not disputed that the fertilizing of acid soil is a waste of fertilizer. Acid soil is a jail for nitrogen and phosphorus, and applying more of the same merely crowds the jail's population. It is interesting that both organic farmers and agribusiness-oriented soil scientists agree that lime holds the key to this jail. (**Note:** A practical minimum application of limestone is 2 tons per acre, since this will not "overlime" any land that shows a need for lime by test — see the "Limestone Requirements" table on page 113 in the Appendix.)

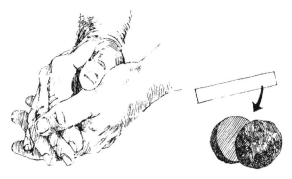

Field Test for Soil pH

1. Using rainwater to moisten the sample, shape it into a ball.
2. Split the sample and slip litmus paper between the halves.
3. Blue shades indicate alkalinity, reds indicate acidity.

CHEMICAL VERSUS ORGANIC

Present-day critics of the organic farmer's arguments claim that condemning the use of chemicals in farming makes no sense, since the soil itself is

Sweet or Sour Soil Test

Add a tablespoon of soil and a few drops of ammonia to ⅔ glass of rainwater. Stir. In two hours clear water indicates sweet, or alkaline, soil; dark indicates a sour, or acidic, sample.

composed of chemical components. A lime application, they argue, is a case in point in that it is a chemical in the same way that phosphorus in superphosphate fertilizers is a chemical, and that both help to make soil productive.

To the organic farmer's claim that "artificial chemical fertilizers" accumulate dangerous quantities of abnormal minerals in the soil, the agribusinessman responds with a grudging acknowledgement that there are growing concentrations of by-product elements resulting from his use of such chemical fertilizers, but that these elements have not been proven to be unhealthy for the soil, nor for humans who consume the product of such soil.

The questions of "good" chemicals versus "bad" ones and that of cumulative harmful minerals are still moot, but the overriding negative results of an overreliance upon chemical agriculture are not. Chemical fertilizers made cheap by government subsidization, and the subsequent *and consequent* increase in the use of fungicides, insecticides, and other chemical poisons are factors responsible for the sorry state of our nation's soils.

Because chemical fertilizers were cheap and effective in producing bumper crops, farmers substituted them for organic fertilizers like livestock manure and green manuring practices (the plowing under of green crops for their nutrient value to the

soil). In turn, the bumper crops led farmers (and absentee investors) to the conclusion that they were no longer bound to the cyclical, self-perpetuating type of small-farm agriculture. They expanded their tillage, consolidated their holdings, discarded profit-inhibiting conservation practices, and undertook intensive cropping on a massive scale. Market demands led to monocropping of hundreds of thousands of acres, and this inevitably resulted in an enormous increase in crop pest populations and crop diseases. Thus the wholesale use of dangerous poisons to rid the crops of pests and diseases can be traced back to cheap chemical fertilizers.

The effects upon the soil of this onslaught of chemicals has been a dangerous lessening of organic matter, and a consequent depletion of mineral nutrients. Without sufficient organic matter to break down, the soil microorganism population has dwindled to unprecedented lows, and because there is less organic matter, the soil's aeration qualities and water-holding powers are enormously diminished. This problem increases exponentially as the "hardening" soils are subjected to drought cycles and leaching. Simultaneously, the farm factory practices of intensive cropping and deforestation leave the fields open to continuing erosion.

Chemical fertilizers are still relatively inexpensive, and they are appealingly easy to apply. They are, therefore, a temptation to the "shoestring" small-scale farmer who finds commercial "organic" alternatives comparatively high-priced and difficult to put on his land. But energy costs are beginning to assert themselves on the market. The energy used in processing chemical fertilizers is considerable, and the steadily increasing prices for this product reflect this fact. Because of the price rises, the coming years will see farmers forced to consider other alternatives, and foremost among them will be those espoused by organic farmers.

COMPOSTING

Composting, an age-old method of making soil from decayed organic matter, is the technique most cherished by adherents of the organic method. In its traditional style, layered piles, it is practical for gardeners, but not for farmers — even those of the small-scale variety. In theory, making compost seems sensible. It does enrich and build topsoil. These are results that any farmer likes, but the problem lies in the scale. Compost has been made on farms having a hundred or more acres of tillage, but reports of these efforts show that the method required large capital outlays for equipment large enough to handle the considerable bulk involved. Most farmers, particularly those of the beginning variety, cannot afford these kinds of outlays.

This is not to say that large-scale composting will always be an impractical alternative for farmers. Given time and the increasing prices of commercial fertilizers, farmers and farm equipment manufacturers may be forced to give priority to alternatives like large-scale composting.

SHEET COMPOSTING

Meanwhile, sheet composting is a viable, soil-enriching practice for almost any farmer. This is the practice of working raw green material into the soil surface, where it decomposes. Most farm harvesting operations leave a certain amount of the crop ("trash") behind, and this could be considered a sheet compost if it were plowed down. Even weed growth, if mowed before the seeding stage, provides one or more welcome layers of sheet compost.

As noted above, sheet composting is a soil-enriching practice, not necessarily a soil-building one. To actually build soil the farmer has but two choices: he must either bring in material to add to his own soil, or he must grow sod. The former involves the costs of the materials themselves (though they are sometimes free), and the costs to handle them (hauling, spreading, etc.). The latter means that the farmer must commit the land to sod for the term of the rotation, buy the seed, sow it, grow the crop, and return it to the soil for decomposition rather than harvesting and seeding the crop. In short, it is root growth that builds soil. Cropping, no matter how "soil-saving" the reputation of the particular crop, is no technique for building soil.

Bringing in material to build up the topsoil layer is, with the exceptions of free manure, sludge, or industrial wastes, an impractical alternative for most farmers. Thus, growing soil-building crops and plowing them under is the better choice. Green manuring farm practices have been used successfully since the time of the Romans — a lesson of history from which we should endeavor to learn something.

GREEN MANURE

Choice of the type of crop for green manuring and the timing for plowing it under are decisions that each farmer has to make for himself. One piece of land may need large quantities of organic material to improve its bulk, and another may simply need more nitrogen. As a general rule, most green manure crops are best plowed under while they are still tender and young. Burying heavy stem growth provides considerable organic bulk, but the soil organisms and bacteria that break this material down tie up valuable nitrogen in the rotting process.

Nitrogen need is an important factor for most farmers in their choice of a green manure crop, so most of them settle on a legume. This unique group of plants provides both abundant organic bulk and takes nitrogen from the air to deposit, or "fix," it in the soil in the form of usable soil nitrates. Alfalfa or sweet clover, for instance, furnishes as much as 40 pounds of nitrogen per acre, red and Ladino clovers 30 pounds per acre, and Lespedeza 20 pounds per acre. It has been estimated that the air above an acre of farmland contains 35,000 tons of nitrogen, and legumes tap this free source of fertilizer at no extra charge to the farmer.

Green manuring *does not* increase the actual mineral content of the soil, but it will make more minerals *available* for use by subsequent cash crops. Mineral availability is increased as the rotting green matter produces various acids (mostly carbonic) that have a solvent effect on the resident soil particles. Once freed from the parent soil particles, the minerals become soluble in the soil's water and thus available for use by a cash crop's plant roots.

ORGANIC FERTILIZERS

For outside sources of the other two fertilizing minerals, potash and phosphorus, most organic farmers who have acreages larger than garden sizes turn to finely crushed rock materials. Rock phosphate and granite dust provide phosphorus and potash, respectively, and they are common examples of rock fertilizers. Greensand, another potash fertilizer, requires no crushing process, because it already has a naturally fine consistency.

All three of these fertilizers rely on their fineness for their solubility in soil waters. Finely ground rock fertilizers release their nutrients at a slower, "steadier" rate than their chemical counterparts. Because they release their minerals slowly, they do not have to be applied as frequently as the instantly soluble chemical fertilizers. These benefits, plus the fact that they are easy to handle and spread in their dry forms, make them good choices for the small-scale farmer.

The rock phosphate of today is different from that sold thirty years ago in that it is much more finely ground. In this state it is composed of a host of trace minerals, and contains 60 to 65 percent calcium phosphate or bone phosphate of lime (depending upon the source). Unlike superphosphates, rock phosphate has not been acid-washed to increase its solubility. It thereby avoids unbalancing the microorganism population of the soil with calcium sulfate (a by-product of the sulphuric acid that is used in chemical processing).

Granite dust can be found at quarries as a waste by-product. This rock material is relatively rich in potash — varying between 3 and 6 percent, depending upon the source. Occasionally some quarries mine granites that are rich in potash feldspars and micas, and they will yield as much as 8 percent potash.

Pricing fertilizers today is a deceiving business. At first reading, most commercial organic fertilizers seem higher in comparative bulk cost than the chemical ones. This disparity soon evens out when frequency of application is taken into account. One might reasonably speculate, also, that the costs for organic fertilizers will decline as the fertilizer industry tools up for greater production.

These organic rock fertilizers require less energy in production processing than do the chemical fertilizers, and this is an important factor in the future of agriculture. We have arrived at our sorry state of agriculture via the road of deceptively cheap energy, and now that this energy is no longer cheap or abundant, we are going to have to retrace our steps back down that road.

At this writing we depend upon fossil fuels for about 96 percent of our energy use. Current estimates are that we will deplete these sources (and nuclear fission sources as well) in one hundred years. Alternative energy sources are being developed — slowly. Unfortunately, the state of our farm soils is such that we cannot afford to wait for this transformation.

Farmers cannot hope to stop the ponderous natural movement of the soil to the sea, but they must curtail their wrongheaded hastening of this process. Tried-and-true farm practices must be readopted, the scale of our agricultural ventures must be cut back, yet the overall acreage devoted to farming increased and the distribution of farm products localized if we are to save our soil and our agricultural heritage.

Glossary of Soil Terms

This is a handy glossary of technical terms that are used by farmers in making soil interpretations or in analyzing a soils map.

AERATION, SOIL. The process by which air and other gases in the soil are renewed.

AGGREGATE, SOIL. A single mass or cluster of soil consisting of many primary (sand, silt, clay) soil particles. Also called a *ped*.

AVAILABLE MOISTURE CAPACITY. The ability of a soil to hold water that will not drain away and that can be used for plant growth. This is usually expressed as a fraction of an inch of water per inch of soil.

AVAILABLE SOIL MOISTURE. Moisture, in a soil, that plants can use. The difference between the

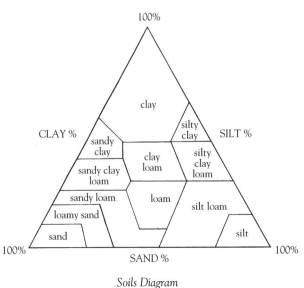

Soils Diagram

moisture content of an adequately drained soil at field capacity and the moisture content when plants wilt is called the "available soil moisture." This definition assumes that the water table is below the reach of the plant roots.

BEDROCK. Solid rock either exposed at the surface or covered with unconsolidated material.

BEDROCK, DEPTH.
   a. Shallow to bedrock. Less than 20 inches to bedrock.
   b. Moderately deep to bedrock. 20 to 40 inches to bedrock.
   c. Deep to bedrock. 40 inches or more to bedrock.

BOULDER. A more or less rounded stone with a diameter of 2 feet or more.

BOULDERY. The term applied to soils that have 3 to 15 percent of their surface covered with boulders.

BOULDERY LAND. A land type that has 15 to 90 percent of its surface covered with boulders.

CLAY.
   a. As a soil separate in the USDA system, the mineral soil particles have to be less than .002 millimeters in diameter.
   b. As a soil textural class, see diagram in Soils section, page 11.

COBBLESTONE. A rounded or partly rounded piece of rock, 3 to 10 inches in diameter.

COBBLY. A soil containing between 15 and 50 percent cobblestones, by volume.

COMPACTION. Packing soil into tight layers.

DEPTH, SOIL. The following classes are used to express soil depth:
   VERY SHALLOW — less than 10 inches deep.
   SHALLOW — 10 to 20 inches deep.
   MODERATELY DEEP — 20 to 40 inches deep.
   DEEP — more than 40 inches deep.

DRAINAGE, NATURAL SOIL. Refers to the condition that existed during the development of the soil, as opposed to altered drainage, which is commonly the result of artificial drainage.

   The following classes are used to express soil drainage:
      EXCESSIVELY DRAINED — water is removed very rapidly from the soil. These are droughty soils and commonly are very shallow or shallow to bedrock, or are very sandy.
      SOMEWHAT EXCESSIVELY DRAINED — water is removed rapidly from the soil. These are fairly droughty soils, and commonly are moderately deep to coarse sand, gravel, or bedrock.
      WELL DRAINED — water is removed from the soil readily but not rapidly. These soils are moisture-retentive but are not wet.
      MODERATELY WELL DRAINED — water is removed from the soil somewhat slowly, resulting in small but significant periods of wetness.
      SOMEWHAT POORLY DRAINED — water is removed from the soil slowly enough to keep it wet for significant periods but not all of the time.
      POORLY DRAINED — water is removed so slowly that the soil is wet for a large part of the time.
      VERY POORLY DRAINED — water is removed so slowly that the water table remains at or near the surface for the greater part of the time. There may also be periods of surface ponding.

DRY DENSITY. The weight per unit volume of a dry soil.

EROSION, SEVERE. The removal by water or wind of the surface soil and subsoil so that the lower subsoil or the parent material is exposed.

FLAGGY. A soil that contains between 15 and 50 percent, by volume, of relatively thin fragments 6 to 15 inches long, of sandstone, limestone, slate, shale, or schist. A single piece is a *flagstone*.

FRACIPAN. A dense, brittle, slowly permeable subsurface layer which occurs 12 to 40 inches below the surface and which may vary in thickness from a few inches to several feet.

GLACIAL TILL. A mixture of materials which may contain boulders, stones, gravel, sand, silt, and clay moved and redeposited by glacial ice with little or no transportation and sorting by water.

GRAVEL. Rounded pebbles between 2 millimeters (.08 inch) and 3 inches in diameter.

GRAVELLY. A soil that contains between 15 and 50 percent gravel, by volume.

HARDPAN. A hardened or cemented slowly permeable soil layer.

HORIZON, SOIL. A layer of soil approximately parallel to the soil surface, with distinct characteristics produced by soil-forming processes.

INFILTRATION. The downward entry of water into the

surface layer of the soil.

LEACHED LAYER. A layer in which soluble constituents have been dissolved and removed by the passage of water through the soil.

MAPPING UNIT. Mapping units are composed of a soil or soils having defined properties. Also included are small areas of other soils that cannot be separated because of the limits imposed by the scale of mapping. A unit may have up to 15 percent inclusions of other soils. On a soil map it is an area enclosed by a soil boundary and identified by a symbol.

PARENT MATERIAL. The original materials from which a soil is formed.

PERMEABILITY, SOIL. The rate at which water will move through a saturated soil.
a. Rapid. A rate of 2.0 to 6.3 inches per hour.
b. Moderate. A rate of 0.63 to 2.0 inches per hour.
c. Slow. A rate of 0.2 to 0.63 inches per hour.
d. Very slow. A rate of less than 0.2 inches per hour.

pH VALUE. A numerical means for designating relative acidity and alkalinity. A pH value of 7.0 indicates precise neutrality; a higher value, alkalinity; a lower value, acidity.

PONDING. The impounding of water on the surface of the ground, caused by slow or very slow infiltration or permeability of the soil, or by the overflow of streams.

PROFILE, SOIL. A vertical section of the soil through all its horizons or layers, from the surface into the parent material.

ROCKINESS. The presence of bedrock exposures within a soil area.

ROCK OUTCROPS (ROCK LEDGES). Solid bedrock exposed at the surface.

ROCKY. Soils with bedrock exposures approximately 100 to 300 feet apart.

ROCKY, EXTREMELY. Soils with bedrock exposures approximately 10 to 30 feet apart. (See note that follows Rocky, Very.)

ROCKY, VERY. Soils with bedrock exposures approximately 30 to 100 feet apart. (**Note:** Extremely rocky and very rocky soils are not separated in mapping. The name used for a mapping unit indicates the condition estimated to be dominant.)

RUNOFF. That portion of the rainfall which does not enter the soil but runs off the surface.

SAND.
a. As a soil separate in the USDA system, individual rock or mineral fragments having diameters ranging from 0.05 millimeters to 2.0 millimeters.
VERY COARSE SAND — 2.0 to 1.0 mm.
COARSE SAND — 1.0 to 0.5 mm.
MEDIUM SAND — 0.5 to 0.25 mm.
FINE SAND — 0.25 to 0.1 mm.
VERY FINE SAND — 0.1 to 0.05 mm.
In most places, sand grains consist chiefly of quartz, but they may be of any mineral composition.
b. As a soil textural class, see diagram in the Soils section, page 11.

SERIES, SOIL. A group of soils having similar kinds of soil layers. The arrangement of these soil layers and their colors, textures, reaction, chemical composition, and other properties are very similar wherever the soil series occurs.

SHEET COMPOSTING. Decomposition of organic material on the soil's surface.

SILT.
a. As a soil separate in the USDA system, individual mineral particles of soil that range in diameter from 0.002 millimeters to 0.05 millimeters.
b. As a textural class, see diagram in the Soils section, page 11.

SLOPE. The rise or fall of the land, usually measured in feet per hundred feet, or percent of slope.

SOLUM. The upper part of a soil profile, above the parent material, in which the processes of soil

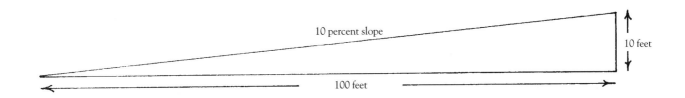

10 percent slope

10 feet

100 feet

formation are active. The solum includes the topsoil and subsoil horizons.

STONE. A rock fragment larger than 10 inches in diameter.

STONY. Soils with stones covering about 0.01 to 0.1 percent of the surface.

STONY, EXTREMELY. Soils with stones covering about 3 to 15 percent of the surface. (See note that follows Stony, Very.)

STONY, VERY. Soils with stones covering about 0.1 to 3 percent of the surface. (**Note:** Extremely stony and very stony soils are not separated in mapping. The name used for a mapping unit indicates the condition estimated to be dominant.)

STRUCTURE, SOIL. The aggregation of soil particles into clumps, peds, or clusters of primary particles. A moist or dry soil may break into pieces of a rather definite shape, such as granules or crumbs, blocks, or plates.

SUBSOIL. Technically, the "B" horizon of a soil with a distinct profile; commonly, that part of the soil profile lying below the surface layer.

TEXTURE, SOIL. The composition or make-up of soil on the basis of the relative proportion of the different soil separates: sand, silt, and clay. The relation between the USDA textural classes and their content of sand, silt, and clay is indicated in the diagram in the Soils section, page 11.

TILTH. A farmer's term meaning loose, mellow soil.

TOPSOIL. The uppermost layer of soil, technically referred to as the "A" horizon. Usually a dark-colored soil or soil material that is commonly used to topdress road banks, parks, gardens, lawns, etc.

TRASH. The organic material left on a field surface after harvesting.

WATER TABLE. The upper limit of the part of the soil or underlying rock material that is wholly saturated with water.

WEATHERING. The physical and chemical disintegration and decomposition of rocks and minerals.

CHAPTER 2

Plants

Without plants our planet would be sterile. They are the foundation of all food chains, and thus the basic consideration of the farmer's growing plans. While a background in botany would be helpful to the farmer, most manage to get by with some fundamental knowledge. Unfortunately, this knowledge is usually more intuitive than scientific or pragmatic, and this self-imposed limitation narrows the farmer's options.

Not all plants are farm crops, but many of them are used in an indirect fashion to produce crops. Plants can be divided into four major divisions: the algae-fungi (*Thallophytes*), mosses (*Bryophytes*), ferns (*Pteridophytes*), and seed plants (*Spermatophytes*). For all intents and purposes the farmer is concerned only with the algae-fungi and seed plants.

The algae-fungi are for the farmer a mixed category of good and bad plants. Algae are of no particular significance to him, but fungi are responsible for many of the blights, rusts, smuts, wilts, and rots that afflict

his crops. On the other hand, fungi also include those bacteria which rot and thereby release nutrients from the organic material in the soil. They also cluster in nodules on the roots of clovers and alfalfa, where they take nitrogen from the air in the soil and convert it to a usable form for other plants. Other kinds of fungi bacteria ferment silage, wine, beer, and cheese. The importance to the farmer of these plants cannot be stressed too much, and much of the daily work he does is aimed at encouraging these fungi by making a medium conducive to their growth and reproduction.

More obvious, but no more important to the farmer, are the seed plants. They make up over half the known species in the plant kingdom, and are considered to be the most complex and functionally specialized of all the plants. Botanists break the seeded plants down into two major categories, the naked-seeded (*Gymnosperms*) and the enclosed-seeded (*Angiosperms*). In large part the former are cone-bearing plants such as pines, firs, spruces, hemlocks, larches, cedar, and cypress, and they are further distinguishable by their growth habit of forming annual concentric rings in their woody trunks. Because they may be considered a farm woodlot crop, they warrant a share of the farmer's attention.

But the enclosed-seeded plants (also known as flowering plants) are the ones with which the farmer has most of his dealings, for they include most of the cultivated farm crops. The last botanical subdivision that most farmers use splits the enclosed-seeded/ flowering plants into two subclasses, *monocotyledons* and *dicotyledons*. Put off by long, five- and six-syllable Latin words, the farmer usually shortens these to *monocots* and *dicots*. The differentiation is made on the basis of the plant's seedling characteristics. Monocots have one initial seedling leaf, and include grass plants such as timothy, brome grass, corn, oats, wheat, rye, barley, bluegrass, and quack (or "witch") grass. Dicots possess two starting leaves, and include the legumes such as alfalfa, medium red clover, Ladino clover, white clover, alsike clover, peas, peanuts, and beans.

Because of their importance to the farmer, four grasses and two legumes were selected from the flowering plants for detailed examination as cash crops in a later section of this book. These six cash crops represent but a minute fraction (six out of 3,000) of the species of flowering plants grown in this country, but in economic significance they dwarf all the others.

FEEDING PLANTS

Plants' nutrient needs, as they are made available by soil, were described in the previous chapter. These needs are *elements* (substances that are irreducible by chemical means). One of the plant's mechanisms for absorbing these elements was described as being the root system, but how this was accomplished was not discussed. Roots take in water and elements through root hairs (with the notable exception of a few plants like blueberries). These hairs are located near the tips of the roots, and they increase the surface area of the root by as much as twenty times. The amount of surface area is important, because it is only through the cell wall and the semipermeable membrane immediately adjacent to the soil that nutrients are allowed to enter the plant. Thus, the more root surface, the more nutrients the plant will get.

Nutrients penetrate the root hair cell wall and membrane only when they are in solution (water and acid are essential), and the movement is by osmosis. Osmosis is one of Nature's ways of equalizing things. Scientists are unable to tell us why this flow occurs, and they are forced to resort to the scientific cliché, "probability based on historical observation." Thus, given a lesser concentration of nutrients inside the root of the plant and greater concentration outside it, the nutrients will move (in solution) into the plant. If the farmer has the flow going the other way, he has a sick plant. Once inside the cell wall and membrane, the nutrients use the same observed phenomenon of osmosis to move to the "water pipes," where they are carried to the leaves by diffusion (yet another phenomenon). The "water pipes" are usually made up of elongated living cells having beveled, wedgelike ends. Their ends are fitted together, and have perforations on their joining surfaces through which protoplasm flows.

Nitrogen, one of the more important nutrients required by plants, is absorbed only through the plant roots (legumes are the outstanding exception, in that they take this element in through their leaves as well). Absence of this nutrient shows itself in a lightening of the greenness of the leaves, and finally in an encroaching yellowness. It is of interest to farmers planting annual-type crops that these plants take up nearly all their requisite nitrogen (and other mineral elements) in their early stages of growth. Spring wheat, for example, has gotten about 85 percent of its nitrogen needs before it is half grown. One of the lessons to be learned from this is that late applications of nitrogenous fertilizers on annuals are wasted. Further, some crops like clover actually begin to return nitrogen to the soil in their later stages (after full bloom), and thereby decrease in nutritional value.

Nutrients soluble in water rise in the plant through the "water pipe" cells in microscopic threads that are continuous from root to leaf. At the top, evaporation acts as a gentle suction that transmits itself through surface tension along the entire length of each thread. In larger plants this tension has been found to exceed 2,000 pounds per square inch — enough to elevate the nutrients and water 150 times higher than the tallest plant known.

Once in the leaves, the nutrients undergo a

remarkably intricate transformation to a glucose or dextrose sugar. The sugar is then used as food or transported as surplus sugar or starch to storage places in the plant. Examples of storage places are: roots (carrots, potatoes, parsnips), stems (celery), stems and leaves (onions, asparagus), leaves (lettuce), flower receptacle (apples, strawberries, pears), seeds (walnuts), buds (Brussels sprouts), ovary or ovary walls (peas, beans, cantaloupes, watermelons, apricots, peaches, tomatoes, pumpkins, squash), and floral parts (broccoli). From these repositories a plant can draw nutrients as needed. It can draw upon both sugar and starch or upon fats and proteins that are built from the sugars and starches.

But what specific uses does the plant make of these nutritive elements? We know that 95 percent of the body mass of a plant is nothing more than carbon, hydrogen, and oxygen in combinations that are supplied only by air and water, but the combinations of the fifteen or so elements that make up the remaining 5 percent of the plant mock the chemist in their precise delicacy and function.

Years of research have revealed that calcium develops leaves, new cells, and protoplasm. Sulphur makes new root hairs and is a part of the complex proteins that the plants use. Meanwhile, root tip and seed growth depend upon phosphorus. Like sulphur, potassium is indispensable as an aid for making proteins, and without this element the tips of the plant would not grow. Silicon is a necessary ingredient in the cellulose structure of the plant, and iron is the activating ingredient for chlorophyll. Chlorophyll, the magic ingredient of photosynthesis, is dependent on magnesium.

FUNCTIONS OF LEAVES

Chemists tell us that they can now make small amounts of chlorophyll, but the cost is exorbitant and the results are uncertain. It remains a mystery how this complicated molecule is formed so readily in a young seedling plant. The first leaf that appears from the seed has this incredible ingredient built in. It is interesting that the formula for chlorophyll ($C_{55}H_{72}N_4O_5Mg$) differs from that of human blood only by the substitution of one atom of magnesium (in plants) for the single atom of iron (in humans).

Chlorophyll is found mostly in leaves, and its function is that of separating the spectrum light rays from the sun. It reflects or transmits yellow and blue rays (hence the green color we perceive), and absorbs the orange and red ones to use in combining hydrogen from water and carbon dioxide from the air. Precisely how this is done is still under investigation, but scientists have labeled the process as *photosynthesis*, and The Equation (basic to all life on this planet) is written:

6 carbon dioxide (CO_2) + 6 water (H_2O) + sunlight + living green cells = 1 sugar ($C_6H_{12}O_6$) + oxygen (O_2)

In performing this sugar-making function, leaves serve as food manufacturing centers. Size and number of leaves can be correlated to the size and type of crop. For example, it takes about thirty to thirty-five average-sized leaves to make a decent McIntosh apple, and about twelve to fifteen to fill an average-sized bunch of seedless grapes with sugar. Viticulturists (wine grape growers) are particularly interested in high sugar content in their grapes, and their centuries-old studies in this leaf-to-fruit relationship would fill many books of this size.

Promotion of growth in a crop's leaves is obviously a good farm practice, but equally important are practices designed to retard leaf growth. Mechanical weed control is often nothing more than the practice of cutting off green leaves of weeds so that the plants exhaust food stored in their roots and thus "starve to death."

Leaves are also mechanically fascinating. It is ironic that we are experiencing such difficulty making an economically viable series of photoelectric cells, when plants do it naturally. Viewed in cross-section, leaves have waxy-surfaced upper and lower skins composed of tightly knit, hexagonal-shaped cells. These skins allow light to pass through to be acted upon by the chlorophyll. The latter is packed into elongated cells (palisade cells) that hang down (inside the leaf) from the upper surface. Filling up the rest of the space between the upper and lower skins are a collection of spongy cells that are interspersed with an infinitely fine network of "water pipes" bringing water from the roots and taking sugar to the storage places. Function in the leaf is so efficient that no cell is more than two cells away from the plumbing.

To work properly, the sponge cells are arranged in a haphazard fashion so that air can circulate through the interior of the leaf. This air is admitted through "muscled" little mouths (*stomata*) that are arranged at a spacing of 30,000 to 50,000 per square centimeter on the bottom layer of the leaf.

All of these leaf functions work so efficiently in the production of food that the plant only uses 25 percent of what it produces for growth; the remaining 75 percent goes to storage. The same may be said about the plant's use of water, for only a small part of that taken in by a plant's roots is retained. All surplus water is evaporated through the leaf stomata. This process is known as *transpiration*.

WATER AND PLANTS

If more water is transpired through the leaves than is absorbed through the roots (the only way a plant gets its water), the plant will wilt. Most farmers know that occasional crop wilting will not cut their yield to any great extent, but if the wilting is prolonged they will sustain heavy losses. The plant's stomata attempt to regulate the rate of transpiration by closing when water is scarce, but sustained drought makes this effort futile.

Historically, yields are cut into more by drought than by too much water. This is discouraging to the farmer working in hot, dry climates, for his crops generally transpire more than do those in wetter climates. The result is that, where water is scarce and expensive, plants require more water to produce a given yield than their counterparts in areas where water is plentiful. Other factors, like soil fertility, disease, and insect damage affect the plant's ability to make efficient use of water. Plants grown in fertile soil use more water than those in infertile soils because they produce more of everything, but they make more efficient use of the water they receive per unit of production.

Different plants require different amounts of water, and this requirement is a strong determinant in what type of crop is best suited to the farmer's soils and climate. Corn, for example, needs only half as much water as soybeans to produce the same amount of dry matter. In general, the deeper the root of the crop, the greater its need for water.

PLANT REPRODUCTION

A plant's water needs are at no time more important than when it starts to reproduce. This phenomenon is done by plants with seed or vegetatively (or both). Natural vegetative reproduction occurs in many ways, such as by tubers (Irish potatoes), roots (sweet potatoes), or stem stolons (bluegrass, clover, strawberries). The Irish potato is an example of a plant that produces both seed and tuber, but the use of potato seed for reproduction is limited to genetic tinkering.

Other forms of man-assisted plant reproduction include cloning, cutting, budding, and grafting. All of these methods are beyond the scope of this book, but they should be thoroughly researched and understood by farmers who intend to grow crops that depend upon them.

Seed reproduction is the most common type of reproduction that the farmer encounters. All seeds contain a dormant or resting embryo plant complete with a resident supply of food (starches), and this "genetic bomb" is usually surrounded by one or more protective seed coverings. The wonder of this self-sufficient, microcosmic package has awed farmers throughout recorded history, and rubbing elbows with today's sophisticated technology has not lessened the farmer's awe. For the farmer, putting seed in the ground is still the most exciting event of the growing season.

Once the seed is exposed to water, proper temperatures, and air containing oxygen, it begins to grow. Light is not essential for germination. The first

Leaf Structure

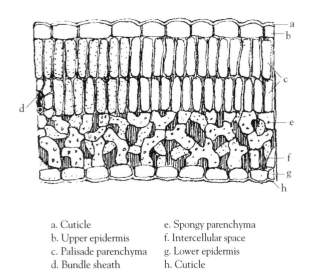

a. Cuticle
b. Upper epidermis
c. Palisade parenchyma
d. Bundle sheath

e. Spongy parenchyma
f. Intercellular space
g. Lower epidermis
h. Cuticle

Monocots

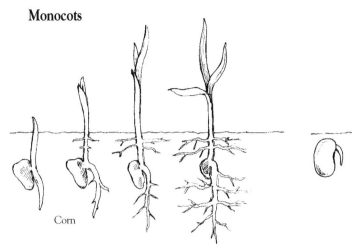

Corn

Dicots

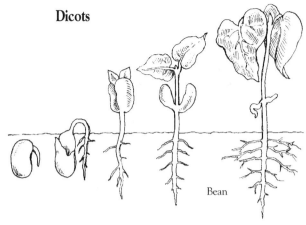

Bean

step usually sees the seed swell as it absorbs water from the soil, and then the mysterious substances we label *enzymes* are released. Enzymes are used by the incipient plant to convert the unusable starches in the seed to soluble sugars. This process is responsible for the farmer's practice of firming his seedbed after seeding by rolling. Vigorous germination is necessary to obtain thrifty plants, and a firm, moist seedbed makes for rapid germination by bringing the seed and soil particles into closer contact.

Planting depths and soil conditions are also important factors involved in the seeding practice. Too deep a planting can result in slow, unhealthy sprouting (or none at all) because of a lack of oxygen or lack of enough resident food supply in the seed. Oxygen deprivation may also occur in soils that are too wet or that are crusted. Seed germination needs vary with different crops, and the farmer must acquaint himself with each. As a general rule, large seeds can be planted deeper than small seeds because they have a greater supply of food, and can afford to take longer to emerge.

As the swelling increases and finally ruptures the covering, the embryonic plant in the seed makes use of the food that is stored to send out a root-tipped stem. This stem turns downward, away from light, and its burgeoning cell growth forces the first seed leaf (leaves) — or the seed itself — upward toward the light. The fashion in which plants actually emerge from the soil varies. Monocots like corn, wheat, oats, timothy, redtop, barley, rye, rice, and sudan grass emerge by (mostly as a result of) cell elongations of the upper part of the plant stem; dicots like alfalfa, sweet clover, medium red clover, buckwheat, beans,

carrots, beets, cucumbers, and watermelon rely for the greater part of this growth on the section of the stem just above the root.

Regardless of the characteristics of emergence, once the seed leaf or leaves are exposed to light, they turn green as chlorophyll is created from protoplasm. The primitive root system expands to meet the nutrient needs of the first leaves and those that follow. (It is interesting that a litmus paper test of these young roots shows the presence of acid. The latter is exuded by the plant to aid in the fast intake of nutrients by making them more soluble.) Monocots like corn rely on the initial root system for this process of sprouting, and then discard it as they develop a second, permanent root system farther up the stem. Since this permanent root system always develops near the soil surface, deep planting in order to promote deep roots is useless. Because deeper roots are desirable in arid areas, corn growers there often resort to trench planting (this practice is called *listing*, and is further described in the Cash Crops chapter), and then to subsequent coverings of the growing plant stem.

Along with soil, seeds are the basis of the farmer's livelihood, and they deserve his careful attention. Poor choice in selecting seeds or careless handling of them once obtained can result in damage that may not show up until harvest time or until it is too late to replant. It is essential for the farmer to know which varieties are best suited to his soils, climate, and practices, and then, even with the reassurance of certification of his seed, to make purity and germination tests of his own before putting the seed in the ground.

The simplest germination test is to lay a moist piece of blotter or absorbent cloth on a kitchen plate, then count out 100 randomly picked seeds onto the surface. (This may try the eyesight when working with small seeds like timothy.) Cover them with a piece of paper, another moist piece of blotter or cloth, and a second plate. Leave in a warm place until germinated (usually two to four days), and then count the sprouted seeds to arrive at a percent of germination.

Space is too limited here to discuss the business (and it has become a business) of seed hybridization. In the past, farmers relied extensively on their own farm-grown seed, but complicated genetic tinkering has relegated this practice to that of a mere occupational curiosity. But for those farmers who grew "pure" strains of a crop, the pragmatic practice of selecting the healthiest plants for seed production was an obvious way to "weed out" undesirable characteristics such as susceptibility to disease. Unfortunately, our nation's seed growers and seed "banks" (national repositories for the safekeeping and propagation of seed strains) are increasingly unconcerned with old-fashioned breeding of healthy, pure strains. Hybridization for increased crop yields and crop-handling characteristics has become the acme of government and private research and development projects. This has had the effect of allowing older, more disease-resistant strains to languish, and sometimes disappear completely. Far too little attention is being paid to breeding present-day plants with some of these older varieties.

To produce seed — for reproduction or for animal consumption — the farmer needs a rudimentary understanding of the flower. Flowers are, from plant to plant, quite different. Corn, squash, and melons, for example, separate their flower functions by having their stamens and pistils in different places on the same plant. Flowers of the legumes have the whole flower spectrum (petals, sepals, stamens, and a pistil), but those of the small grains lack petals and sepals. Finally, there are those flowers that lack either a pistil or stamens, such as strawberries or asparagus.

It is the *pistil* (female part) and the *stamen* (male part) that the plant depends on for fertilization. Pollinization involves the transfer of pollen from the stamen to the pistil. Self-pollinating plants like wheat, oats, barley, and soybeans rely on wind and gravity for this transfer, while alfalfa, medium red clover, alsike, and white clover are examples of those that depend on bees and other insects for cross-pollination. In the natural design of things, wind-pollinated flowers have no need for the showy petals or fragrant odors used to attract the helpful insects that are so necessary to cross-pollinated varieties.

Very shortly after the pollen touches the pistil, it descends through a tube in the pistil into the ovary, where it and the egg cells unite to produce the new embryo plant and its endosperm. In some plants the fertilization can take place in minutes, while in others it may take hours. Their function fulfilled, the petals and stamen wither away, and the ovary continues to harbor its maturing seeds. Four to nine weeks are usually required before the seed is mature, but, depending upon the crop, the farmer may harvest them prematurely.

PLANT PROBLEMS

One cynic described cropping as "one penultimate disaster leading to another." There are times for a farmer when, in the throes of his eternal battle with weeds, disease, insects, and other pests, he tends to agree. It seems unfair, for example, that a single burdock plant can produce 20,000 seeds, and a large purslane plant as many as a million, while two healthy ears of corn can only yield a few hundred. And there is not a farmer living who hasn't seriously wondered how he could develop a market for his quack grass.

Weeds are usually defined as plants growing where they are not wanted. Hard as it is for the farmer to stomach, they are occasionally handy to have around, as when, for one reason or another, a field has to be left open. In this case weeds make a "poor man's groundcover" that holds the soil and nutrients back on the land; and, when eventually plowed under, they provide welcome organic material to the soil.

Nevertheless, weeds are, in general, more injurious than helpful. Small-scale farmers operating on narrow profit-to-yield margins know how these unwanted plants cut into their yields and crop quality, and they begrudge time spent on weed control that they might spend more profitably. As the farmer soon learns, weed control is a profound description, for one does not eradicate them. The exception to this rule is poisonous weeds like black nightshade, jimson weed, poison hemlock, locoweed, poison ivy, poison oak, poison parsnip, death camas, dogbane, cocklebur, and rough pigweed.

Weed control begins with prevention. Planting dirty seed is a profitless pursuit, yet farmers frequently get so harried in the spring rush that they neglect to inspect and, if necessary, clean their seed. Other prevention techniques the farmer should employ are to be fussy about buying hay or grains for livestock feed or straw for mulch that shows significant quantities of weed seeds or seedheads. It is more difficult to avoid buying manure that is impregnated with weed seed, and the farmer is advised to buy only from another farmer who shares his attitude toward weeds.

If the farmer can narrow his weed control concerns down to keeping his resident weeds from multiplying, he is ahead of the game. Diminishing their numbers is his goal. The most common method used today to control resident weeds is to spray with herbicides like 2,4-D — despite the serious questions raised concerning their environmental and health effects. Because the small-scale farmer is working with small acreages, old-fashioned, mechanical controls are both feasible and desirable alternatives for him.

To make use of these controls he should first identify his weeds and then understand their growth habits. There are three types of weeds: *annuals* like lamb's-quarters, pigweed, purslane, wild mustard, crabgrass, chickweed, ragweed, downy brome grass, green foxtail, Indian mustard, dodder, yellow foxtail, and wild oats; *biennials* like wild parsnip, burdock, bull thistle, mullein, wild carrot, and poison hemlock; and *perennials* like quack grass, dandelion, narrow-leaved plantain, milkweed, Johnson grass, field bindweed, yellow dock, sheep sorrel, daisy, broad-leaved plantain, wild onion, and Canada thistle.

Annuals grow from seed each and every year. They divide into summer and winter types. The summer annual starts its cycle in the spring and produces seed in the fall, while the winter annual germinates in the fall and produces a rosette-shaped clump of leaves before winter. It then completes its cycle to seed the following summer. Both summer and winter annuals have shallow root systems that lend themselves to easy elimination by cultivation — particularly during early stages of growth. They may also be destroyed by mowing at blossoming time. These are the easiest weeds to destroy.

Biennials start their cycle in the spring of the first year, form a leafy rosette before winter, and resume with stems, flowers, and seeds the following year. Clearly, mowing these weeds the first year will not eliminate them, and cultivation, if done too late in the summer, will not get at their deep taproots. The most effective program is to mow, cutting them at blossom time of the second year. If they are mixed with annuals, this program should be extended to two years.

Perennial weeds are the worst of the lot. They reproduce both by seed and by vegetative means (creeping stems, tubers, and bulbs), and they have extensive root systems. Control programs which rely on mowing alone are not effective, but if the plants are not allowed to produce any (or very few) leaves above the ground, they will finally use up their root-stored foods and die.

Frequent cultivation is the most effective means of control for larger acreages, and small, sporadic outbreaks should be controlled by smothering with mulch, manure, or black plastic, or by individual hand-digging.

No plant is immune to disease or the depredations of pests, and, depending upon the use the farmer plans to make of his crop, his job is to cultivate healthy plants by whatever means he finds to be ecologically and economically sound. This is a burdensome responsibility that deserves the farmer's continuing scrutiny and study. Whenever plant diseases or pests are discussed, the question of chemical controls is bound to come up. Advocates and opponents of chemical sprays are adamant in their positions — positions that often ignore scientific or historic evidence, or are unrealistically idealistic.

Organic farmers, for example, often appear so zealous in opposition to the use of chemicals that they are blind to and intolerant of an economic necessity that may force a small farmer to spray in order to save a threatened crop (or even his whole farm). Usually one will find that the most righteous of these zealots can afford the purity of his arguments by virtue of his personal noncommitment. It is one thing to philosophize abstractly about growing a crop, but quite another to sink one's whole farming future in it.

On the other hand, there is at large in the farm populations of this country an ignorant, uncaring lot of chemical users who would, if unchecked, denude our planet with their irresponsible use of sprays and dusts. Unfortunately they are in the vocal majority, and they are bankrolled by a lucre-ridden chemical industry. The latter has grown fat on the dangerous practices of monocropping and the other malpractices incident to large-scale agriculture.

As a nation, we are just becoming aware of the environmental havoc that irresponsible use of chemicals has wrought, yet their use continues to grow. Farmers and chemical corporations are not alone in this blame. Consumers who have been brainwashed into buying-acceptance of processed foods, blemishless apples, and glue-filled ice cream deserve their share of acrimony. An organic farmer who fifteen years ago invested in planting an apple orchard finds today that his blemished, unsprayed apples are virtually unmarketable, and that fresh apple consumption by the buying public decreases alarmingly each year (since 1910 apple consumption per person has dropped 70 percent, and the consumption of food coloring in phony foods is up 995 percent). To survive, the apple grower is forced to spray.

Small-scale agriculture that is based on sound husbandry practices has little or no need for chemical controls, and they should *never* be used as a standard operating procedure. In making his decision, the small-scale farmer needs to remember that yields are important, but not so much so that they warrant risking the health of his soils or the quality of his crops. Poisons in pesticides are often slow to break down, and their effect on field-dependent wildlife and other ecologic food chains has proven significant enough to deserve legal prohibitions and closer regulatory control. Their potential harm to humans is still under study.

The scale of the small farmer's operation is such that he can (and should) raise healthier, more thrifty, more vigorous crops. He can do this because he has fewer of them to deal with, and can therefore devote more care to each plant. As a general rule (there are exceptions) diseases and insects seek out, and have the greatest impact on, the weaker plants. Prevention costs are always less than remedial ones, and because the organic farmer may actively want to avoid spraying to prevent diseases or fungi growths, he should redouble his research into resistant varieties of crops before planting. He should also take extra pains to have his soil in tip-top shape, and he should tour the perimeter of his fields to eliminate potential host plants. Lastly, he should plan his crop rotations carefully to avoid successive plantings, which encourage accumulations of specialist parasites.

If, despite these precautions, the farmer's crop is infected with a disease, he should immediately determine the extent of the infection. Culling out the infected plants may prove to be the cheapest control, particularly if the infection is localized. Severe infestations or infections are another matter, and this is the point at which the caring farmer must make a difficult choice.

These same kinds of decisions must also be made when faced with animal encroachments. Often, however, one can make some of these choices before putting seed in the ground. It is, for example, foolhardy to plant beans in areas where deer are thick and state compensation for wildlife damage is minimal or nonexistent. Woodchucks, birds, rabbits, and a variety of the rodent population can and do make enormous inroads into fields, and the farmer must choose between systematic eradication (for this he can select from a plethora of deadly traps, baits, gases, poisons, and bombs), or he can sustain the losses with occasional efforts at control.

CHAPTER 3

Farm Machinery

Farm machinery is run by power, and, because the concept originated with the Farm, we measure power in terms of horsepower, or number of horsepower. A horsepower is defined as a force equal to raising 33,000 pounds one foot in one minute.

This compulsion to standardize is always looked at askance by the old-time farmer. He knows from experience that any decent farm horse could do all that the engineer's definition demands, and then "make change at the end of the furrow."

TRACTORS

This writer agrees with the old-timer but will bypass a discussion of draft animals as a source of power for farm machinery. For farmers with the setup (barns, pasturage, etc.) and the skills necessary to care for, instruct, and work their draft animals, they can be one of the best power source alternatives available. Certainly horses, mules, and oxen are not fossil fuel

consumers, nor do they pollute the atmosphere. The reason for passing over them here is that most newcomers to farming will have neither the requisite skills nor the physical plant to handle draft animals, and, because of the pressures inherent in undertaking a new farm venture, they will not have the time to acquire these skills or facilities.

Tractors are a less romantic but necessary alternative for the new farmer. Now ubiquitous in farm country, they have all but completely replaced draft animals as power sources. This is not the place to review the fascinating historical development of these machines, but the changes wrought on them were a slow, pragmatically dictated progression of events. Many of the tractors still found in farm country are only "one step up" from the horse in that their function is limited to simply pulling machinery that was originally designed for the horse.

A banner year for tractors was 1947, when manufacturers added an independent, built-in power transmission device. This power take-off (now simply called a PTO) allowed the tractor to power an attached piece of farm equipment while the tractor was in motion. Included with the PTO was an additional power transmission unit called a belt pulley. This unit was belted to the driven unit, but it could only be used when the tractor was stationary.

Today PTOs can be found on every size and power of tractor, from the 5-horsepower walk-behind garden models to the 140-horsepower behemoths now in use on big megamechanical farms in the Midwest. The belt pulley, on the other hand, has been left off almost all the newest tractors.

The small-scale farmer does not need large quantities of horsepower to handle his operation, and he should confine his tractor shopping to the smaller ones having a horsepower efficiency somewhere in the lower third of the available range. Size of acreage, physical layout of the farm and its soils, and kind(s) of crop(s) grown are all factors to be considered when buying a machine. Farmers, for example, who are working very small areas (say three to five acres) with intensive gardening techniques, have no use for anything larger than a small garden tractor, and they might even get by with one of the smaller walk-behind units.

Small Tractor

Farms on the scale of 10 to 100 or so acres of moderately easy-to-work soils, can readily be worked with tractors of the 20- to 40-horsepower range. (Draft and power requirements for various crop machines can be found in the Appendix.) Since many of the older tractors appearing in the late 1940s and in the 1950s were built to last (and are plentiful on

the market), the newcomer to farming should not confine his shopping to newer machines.

Foremost among these older, rubber-mounted tractors are: International Harvester's Farmall Model H (or Model M), John Deere's Model B (or Model A), Allis Chalmers' Model W30, Case's Model SC, Oliver's Row Crop Model 70, and Ford's Model 9N. Tractor connoisseurs may take exception to inclusions or exclusions from this listing, but they will not quibble about the deserved popularity of each of these particular machines. Excepting the Case SC and the Ford 9N, these tractors were factory-equipped with narrow (row-crop) front ends, and they varied in their methods of hitching implements, from the primitive one-point hitch of the Farmall H to the more modern three-point hitch of the Ford 9N.

These idiosyncratic factors, plus sophistication of hydraulic systems, should be taken into consideration by the buyer. Crop choice, type of soils, terrain, and kinds of implements also bear directly on the selection, and each farmer has to decide for himself which tractor best fits his needs. The more complicated the choice of mechanical systems becomes, the more expensive it is to purchase and to maintain.

The economics of owning and maintaining farm machinery is also more complicated than one would first suspect. After buying the machine (and perhaps paying interest on a loan), one must consider the depreciation, repair costs, cost to house it, and any costs of taxes and insurance. Detailed tables for estimating these costs may be found in the Appendix, but a rule-of-thumb estimate of overhead can be arrived at by:

1. Dividing the initial purchase price (delivered) by eight to get the average cost per year, then
2. Dividing this yearly cost by the average days, hours, and acres to get the respective overhead costs.
(**Example:** A machine costing $1,500 and used 20 ten-hour days on a tillage of 100 acres would have an average overhead cost of $187.50 per year, $9.38 per day, $.094 per hour, or $1.88 per acre.)

On first look, the new farmer can be overwhelmed by the seeming complexity of farm machinery, but this awe disappears as he works with each piece.

Fortunately, farm equipment is designed for utility rather than style, and, with the exception of necessary safety shields, the working "guts" of most machinery are open and available for inspection, analysis, and, when necessary, repair.

OTHER MACHINERY

Farm implements are an esthetically pleasing study in simple utility. Effect follows cause in a chain of remorselessly logical mechanical relationships. One does not have to submerge himself in esoteric engineering data to understand the simple working principles of most farm machines. From the power train of the tractor to the working surface of the tool, there are nine common mechanical components with which the new farmer should acquaint himself. They are: levers, shafts, U-joints, compression clutches, bearings, gears, chain drives, sheaves, and V-belts.

Levers are familiar to most of us from high school physics courses, and they are used extensively on farm equipment. This is particularly true of older equipment that was in use before the popularity of hydraulic systems. The leverage principle is, of course, still employed through farm engineering mechanics, but it is now mostly confined to machine components levering other machine components. Modern-day agricultural engineers seem singularly devoted to eliminating those levers that allowed the farmer to apply his tools to the soil (i.e., lifting and lowering heavy equipment).

Despite its disfavor with designers, the farmer-operated lever has two significant advantages over its hydraulic system successor. It had fewer moving parts, thus requiring less energy, and its functional simplicity resulted in less breakdown time — the bugaboo of all farmers.

One of the more reliable mechanical components of farm machinery is the power shaft. Barring metal fatigue, these shafts rarely give trouble. This reliability is probably due to the other three mechanical components that are commonly associated with them: bearings, couplings, and compression (slip) clutches.

Power shafts are designed to transmit *torque* (rotational pressure) at various speeds. The higher the speed of rotation, the greater the need for straightness of and balance in the shaft to prevent distortion due to vibration. Bearings solve part of this problem by providing guiding support for the

shaft and, of course, reducing friction to a minimum. The two basic kinds of bearings are *journal* and *rolling contact*. Journal bearings are simply metal (occasionally wood) sleeves in which the shafts turn. Depending upon the speed of shaft rotation, these journals may or may not be lubricated. By contrast, rolling contact bearings are always lubricated, usually with a heavy grease. Rolling contact bearings have variously shaped rollers, but these rollers are always assembled sandwiched between an inner and an outer "race."

As shafts need to be rigidly steady, they also need to be able to deliver their torque loads flexibly at varying angles to the driving source. This flexibility is accomplished with various kinds of couplings of which the most common is the universal (or U-) joint. This joint provides for the greatest misalignment (15°) of the two shaft segments. In construction it is composed of two Y-shaped joints (one at each end of the shafts to be coupled) that are joined, through bearings, to a metal cross. The Y-shaped joints, swinging at right angles to one another, then allow the transmission of power from one angle to another.

U-Joint

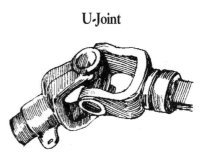

Inherent stresses within the machinery itself are one thing, but day-to-day working hazards are quite another. Shafts that are suddenly stopped, as when an implement plugs up or jams, would be reduced to scrap metal if it were not for ingenious devices like the compression (slip) clutch. Mounted in the middle of a power shaft, it is a kind of angle-jawed coupling that is held in a meshed position by a spring having a set torsional value. When this value is exceeded, such as when the machinery is jammed, the angled jaws disengage, and thus prevent the shaft from assuming the shape of a pretzel or the gears of the driving unit from being stripped.

Perhaps the most fascinating and best designed components of most farm equipment are the gears. When used properly, these toothed wheels can multiply shaft speeds or torques or reduce them; they can transmit power around corners or translate it into differing speeds (differentials); they can even reverse the rotation of two shafts. Boasting a mere one percent power loss per gear engagement, they are among the most efficient of power translators.

Gears come in assorted shapes, sizes, and tooth arrangements, of which the principle ones are the spur, bevel, worm, and rack-and-pinion. When meshed in single sets these gears have a "gear ratio" that is a statement of the relative input and output speeds. The latter is determined by the simple expedient of counting the number of teeth on both the driven and the driving gear. For example, a ten-tooth gear driving a twenty-tooth gear is said to have a two-to-one gear ratio.

Gears are typically used to transmit power over short distances, but when the center-to-center distance between two shafts is large, engineers have reverted to the more economical steel chain drives or to the V-belt and sheaves. Of the various chains used on agricultural machinery, the roller chain is the most popular owing to its relative efficiency (approximately 98 percent). The V-belts do not pretend to this efficiency (at best 95 to 96 percent) because they depend upon frictional forces, but their ability to absorb shock and to act as a mechanical safety factor makes them desirable. This versatility makes them particularly useful on combines.

COMBINES

Of all the farm implements the combine perhaps needs the most reliable power transmission devices. Its multiple functions (mowing, threshing, winnowing, cleaning, bagging, and windrowing) involve nearly every kind of mechanical contrivance found in farm machinery, and it operates under stresses and shocks that challenge the best of engineers.

Self-propelled combines are the choice for today's larger acreages, but the small-scale farmer can still find workable, smaller, pull-behind units — those that were in common use in the 1950s. Powered from the tractor's PTO, these older combines process grain and seed of any kind, and they are, as a whole, easier to service and maintain.

The functional elements of the combine can be divided into five major sections: header, thresher, straw conveyers, cleaners, and bagging area. The header section includes the mowing and delivery machinery that brings the unthreshed grain or seed plus its straw to the threshing cylinders. To push the straw back onto the platform, the reel is coordinated with the combine's wheel speed by means of a

Self-Propelled Combine

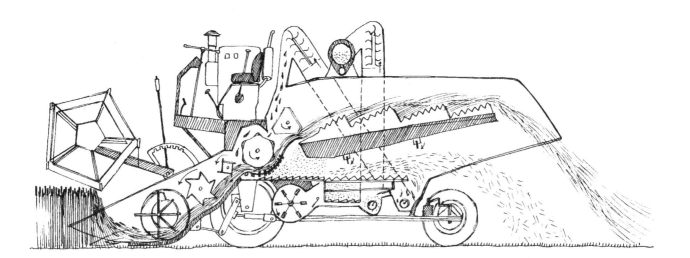

combination of V-belt, sheaves, a chain drive, and sprockets. Older models of combines conveyed the cut crop up the length of the header with canvas aprons. The canvas (later replaced by auger tables) was particularly suited to the gentle handling of delicate crops like beans, but it was prone to backfeeding under the canvas table. Coarse-stemmed crops like sunflowers require special adaptations to the header and the threshing section.

Once into the threshing chamber, the crop is subjected to repeated beatings by adjustable barred cylinders. The bars are usually angled across the cylinder, and may be metal or rubber, depending upon the type of crop. Shattering the seed from the head of the crop is dependent upon the force and frequency of impact from the cylinder bars. The newer combines show better results on some crops (like crimson clover) than the older ones because they offer a range of cylinder speeds.

Everything that goes into the threshing chamber of the combine is then kicked out onto the straw racks. The threshed material is shaken and tossed as the straw raddles push the straw up and out of the combine. Sifting through the straw raddles, the grain or seed falls onto the cleaning screens or cleaning shoe, where it is subjected to a regulated, but constant, air blast emanating from a metal-vaned fan (turning at an average rpm of 500).

The cleaning procedure is usually a combination of agitation and air separation. Agitation takes place on the surface of two adjustable screens that are racked one above the other and in an action that sees them oscillating at about 300 cycles per minute. As the crop is sifted through the screens it drops into a small, horizontal auger table and is then delivered to the bagging platform (or bin) by elevator-type chain-mounted paddles. Grain and chaff that do not find their way through the screens are recycled back to the threshing chamber for recleaning.

Some of the PTO-driven, pull-behind combine models and types are more suited to the small-scale operation than others. Because most small farmers work with less horsepower, it is prudent to restrict the size (width of header) to 6 feet or less.

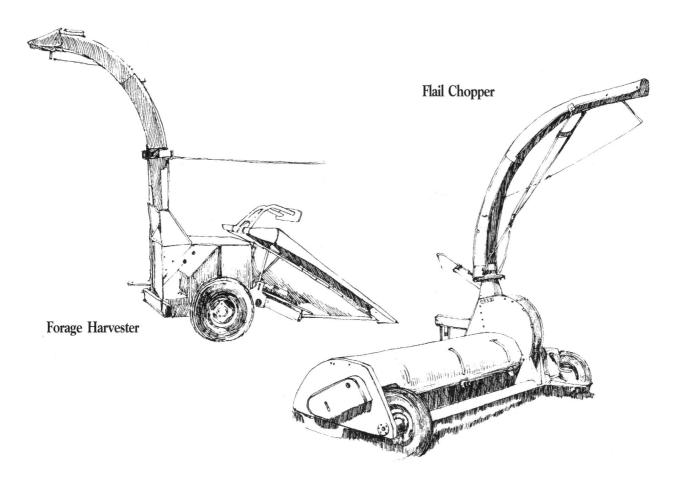

Flail Chopper

Forage Harvester

FORAGE HARVESTERS & FLAIL CHOPPERS

Forage harvesters and flail choppers are recent additions to farm equipment lines, a fact that can be allied with the increasing use of economical trench silos. These machines cut or chop standing forage crops and load them into a wagon or truck in the field. They are commonly used on row crops like sorgo or corn, or with hay crops (green, dried, or partially dried), and the processed crop is typically fed to livestock directly after processing, or is stored as silage for later feeding.

The functional elements of the forage harvester differ quite radically from those of the flail chopper in that the harvester has four separate operations compared to the simpler two-phase operation of the flail chopper. A hay pickup, cutter bar, or row-crop cutting knife leads off the forage harvester process (interchangeable heads), and the crop is then force-fed with feed rolls into the cutting knives. These knives are arranged either cylindrically or radially to cut with a shearing action against a stationary bar — like a large lawnmower. Finally, an impeller blower delivers the processed crop to the truck or wagon that is driven alongside.

The flail chopper's more rudimentary functions evolved from the hammer-mill–type stalk shredder, and it is best used on green or partially dried field crops. Cutting into 2- to 10-inch lengths, it depends upon the impact of its chisel-pointed, hook-shaped knives for both its cutting function and for its delivery operation. Though it is most often used on standing crops, the flail chopper can also be used on windrowed hay because of the suction-pickup action of the whirling knives.

Both the flail chopper and the forage harvester are PTO-driven, and most models are suitable for use with the 20- to 40-horsepower tractors. Small-scale farmers, however, will probably find the flail chopper more versatile and freer of jamming problems.

BALERS

A baler is key to any small-scale haying enterprise. It is designed to compress and tie hay or straw into rectangular or round bales so that is may be more easily handled and stored. While loose hay or straw may still play a part in the self-contained small farm situation, a baler is essential if the grass farmer intends to market his crop. Today's balers come in sizes varying from the small PTO-driven (or independent-engine-driven) models that produce 17" x 36" rectangular bales, to the new machines that turn out round bales the size of a small house.

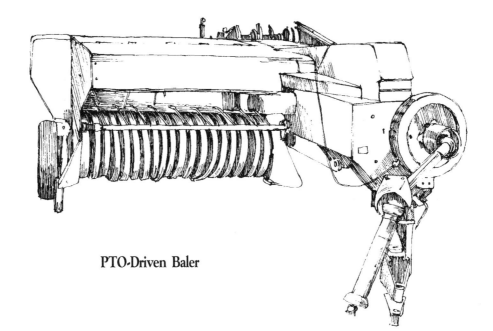

PTO-Driven Baler

The baler has four functional elements: pickups, bale chamber feeding, compression, and tying. The pickup reel is usually made up of spring steel teeth that lift the mown hay or straw over the gapped stripper plates and deliver it to the bale chamber feeder. On most balers the bale chamber feeder is a large cross auger that pushes measured charges of hay into the bale chamber during the intervals when the packer fingers or ram are withdrawn. Once a charge of hay is introduced into the bale chamber, it is packed by the thrust of a ram, plunger, or packer fingers (depending upon the manufacturer), and it is made ready for tying by a final shearing by a knife mounted on the ram or plunger.

The tying procedure differs from baler to baler. A "one-cycled" baler, for example, requires that the needles be introduced, the knots tied, and the needles withdrawn before the plunger compresses the next hay charge. Though slower, this method produces a tighter bale, because the plunger holds the bale in compression while the needles are introduced. The multicycle baler, by contrast, offers greater baling capacity and a simpler mechanical function.

Two parallel binds are typical of both wire and twine balers, but the methods employed to tie them are quite different — wire is fastened by twisting, while twine is tied by a double overhand knot. Both methods use needles to encircle the bale and place the twine or wire in position for the final fastening.

Needles for baler twine come in both open- and closed-end types. The open-end type picks up the twine strand as it starts its stroke, and the closed-end is threaded through its tubular length. Actual knot-tying is done by a knotter bill hook, which is a rotating, slant-jawed device that resembles a duck's bill in profile. In tying a knot it turns once to take a bight of the double strand around its jaws, and then takes the bitter ends in its jaws before withdrawing. This results in the knot described above.

When working with rank, heavy hay, a baler can require considerable peak loads of horsepower. In these circumstances an underpowered tractor can take on the aspect of a dog that is being wagged by its tail, but tractors of the 30- to 40-horsepower range will suffice, when operated with some common sense.

RAKES

The dump rake was a considerable step up in hay-handling from the old-fashioned "bull" rake (an oversized garden rake), and that probably accounts for its continuing use. Long, curved spring tines that are rigidly mounted on a rotatable bar allow the dump rake to be emptied as it crosses the windrow line. When triggered, a cleverly contrived cam that is mounted on the wheel lifts the tines and then drops them to continue their gathering task. The dump rake was traditionally used to gather loose hay, but can be used to windrow for baling.

Sweeping hay into windrows was the next significant advancement in rakes, and the cylindrical-reel side-delivery type made its first appearance just

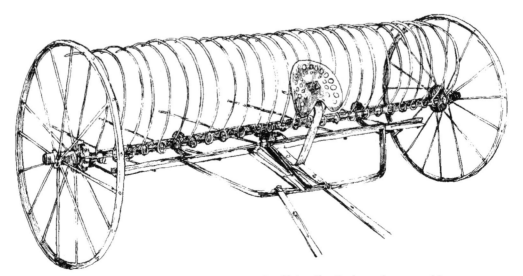

An old-time "hoss" rake can be converted for tractor use.

before World War I. It has remained virtually unchanged to the present day. But the sweeping, side-delivery principle saw one more useful modification immediately after World War II. This came with the "California" wheel rake, which was a wheel-mounted series of floating, overlapping wheels that were tipped with short, feathered spring teeth. The advantage of the sweeping, side-delivery method of windrowing hay (particularly with the California rake) is that it tends to fluff the crop, thus facilitating drying.

Horsepower is not a major factor with any of the rakes cited above, and the small-scale farmer will be smart to acquire the right tool that fits both his budget and the planned extent of his haying enterprise.

MOWERS

Harvesting hay (or any other standing crop) begins with mowing, and the farmer need consider only two alternatives, the rotary mower and the mower cutter bar. The rotary mower is a relatively new addition to farm equipment lines, and has met with considerable

Side-Mount Mower

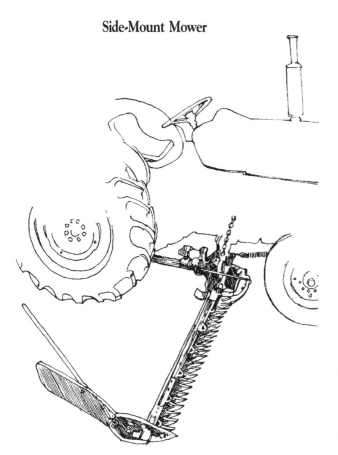

popularity with farmers having specialty crops or acreages and crops on which the rotary mowers can be profitably ganged.

By far the most popular type of mower, the cutter bar, cuts a 5- to 7-foot-wide swath, and it can be mounted in front of the rear tire or on the drawbar, or it can be trailed. It is PTO-driven and cuts by shearing the stems between the reciprocating knife sections and the stationary guards. The knives and guards are mounted on a bar that can be lifted (either hydraulically or by means of a lever) to clear obstructions, or the lift can be exaggerated for transport. In the event that the bar does encounter an immovable obstruction, it is hinged with a breakaway device that is designed to prevent severe damage to the pitmans driving rod or to the mower gearbox.

There are obvious disadvantages to the cutter-bar-type mowing machine. For example, the reciprocating motion of the knives sets up a fatigue-inducing vibration (despite counterbalancing), and the cutter bar is prone to tip-loading in heavy or rank stands of grass. The drawbacks notwithstanding, cutter bars remain the workhorses of the haying operation because they perform their job with the least amount of maintenance.

Small-scale farmers will find that using a cutter-bar-type mower effectively is more a matter of keeping tractor speed down and knives sharp than it is of horsepower. Most mowers are estimated to need 1 horsepower per foot of cutter bar length, and since most of these machines are geared up to about 900 rpm with a forward travel rating of 3 inches per stroke, this means that they should cut efficiently at speeds up to 5 mph.

SEEDERS AND PLANTERS

Seeding and planting machines meter out seed and distribute them at uniform intervals and depths, and, in some cases, compact the soil over them. In the case of seed that can be broadcast, it may simply be done by hand, but this traditional method requires considerable practice in order to get a uniform distribution. The hand-cranked spinner spreader with an agitator and adjustable seed hole is still used, and is quite practical for the small farmer. Bigger tractor-mounted models that are PTO-powered use the same functional principles at considerably greater cost, and

Hand-Cranked "Belly-Type" Seeder

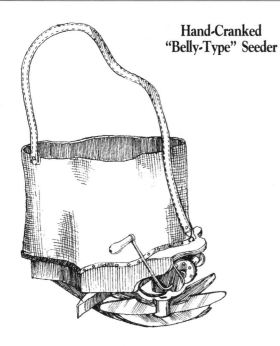

corn planters, potato planters, or peanut planters. These machines have devices like disks or sweeps to open the furrow; fluted or horizontal seed-cell-type metering mechanisms to issue the seed at uniform rates; reversed disks or drag chains to close the furrow; and press wheels to compact the newly closed furrow.

Since many planters are trailed, wheel-driven units, horsepower is not a primary consideration. Specialized planting equipment, such as that used for potatoes, peanuts, or cotton, or for transplanting, should be considered by the small-scale farmer only if he intends to work a large enough acreage to warrant its purchase.

SPREADERS

As with seeds and planters, the functional problem of spreading lime or fertilizers is metering the flow. If the material to be spread were of the dry, granular consistency of sand it would be a simple matter to trail a wheeled hopper unit with holes in the bottom that were sized to meet the speed at which the hopper was pulled. The 8- to 12-foot-wide hopper fertilizer and lime spreaders are precisely this kind of machine. Because of the varying consistencies of materials to be spread (lime, rock phosphate, greensand, etc.), the holes in the bottom of these spreaders are adjustable, and an axle-driven agitator is used to keep

show no better coverage than can be accomplished with the hand-cranked type.

Planting, however, is a different matter. In many instances the farmer who broadcasts his seed may rely on the spatter of raindrops to cover the seed, or he might, at most, chain-drag the newly sown field. But seed that must be planted at a specific depth in the soil is best planted by machines like seed drills,

Wheel-Driven Manure Spreader

the material flowing. This simple tool is a particularly desirable piece of equipment for small farmers who are persuaded to use the organic line of fertilizers.

The organic farmer who raises livestock will also find the manure spreader a necessary tool. Designed to transport as well as evenly distribute manure, the spreader is stoutly made with metal and wood, and is mounted on rubber. Powered by PTO (older models were entirely wheel-driven), a set of conveyor chains (which are joined by angle iron crossbars) feeds the manure back to a double set of beaters. The beaters are mounted one set above the other at the tail end of the spreader, and they are variously composed of spikes, augers, or paddles, depending upon the manufacturer.

Most manure spreaders employ wood floors because of the wear factor inherent with the conveyor mechanism. They present a rugged construction of wood-lined sideboards, and rims that can withstand rough usage. When this heavy construction is combined with the load weight, the overall power requirement is a factor to consider. Small-scale farmers having 30-horsepower tractors should restrict the size of manure spreader they intend to use to those of the 60- to 80-bushel range.

CULTIVATORS AND DISK TOOLS

The extensive use of herbicides has, at this writing, introduced a lot of cultivators onto the used equipment market. Organic farmers who will not use herbicides therefore can usually acquire these machines at a reasonable cost. Before the advent of chemical weed killers, cultivation was the primary method of weed control. Mulching was another, but was not used on extensive acreages. Cultivation was done with gangs of disks, sweeps, shovels, and listers that were either mounted on the tractor or trailed.

One can readily see the scope of the work cultivators were expected to do. Some of the sweeps and shovels were devoted to control of weeds between the rows, a function that was accomplished by uprooting the weed and leaving it on the soil surface to die, or by cutting the weed stem off just below the soil surface. Disks were employed for in-row weed control. Throwing dirt in toward the row, the cultivator smothered the weeds.

But weed control is not confined to the row-type cultivators. Spring-tooth harrows have served farmers since the middle of the last century to control quack grass. Having few moving parts, they have proven

Pull-Behind Cultivators

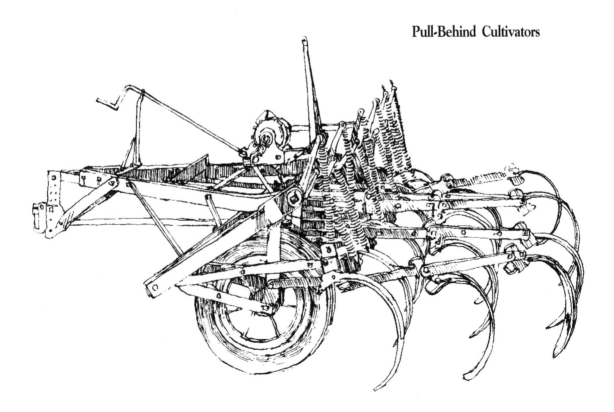

Listers and Hillers

Furrowers/Hillers

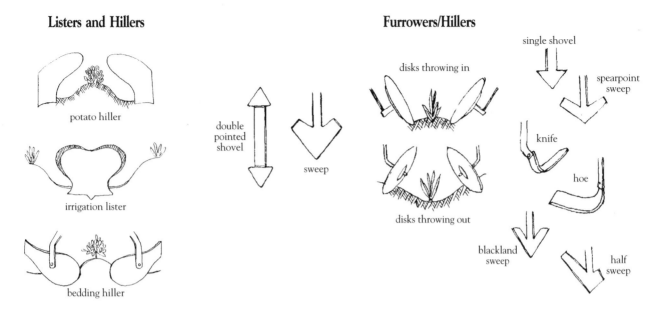

potato hiller

irrigation lister

bedding hiller

double pointed shovel

sweep

disks throwing in

disks throwing out

single shovel

spearpoint sweep

knife

hoe

blackland sweep

half sweep

relatively trouble-free, and are still used on newly cleared fields to loosen soil, break clods, expose root segments, and loosen seedbeds that have been compacted by rain.

In use, three or four separate spring-tooth harrow sections, each 3 to 5 feet in width, typically are ganged on a common drawbar, and trailed behind a tractor. Depth of penetration is controlled by raising or lowering the spring teeth by means of a rotating tooth bar. This tooth bar, in turn, is manipulated with a ratcheted lever on the harrow section, or by a trip-rope that can be activated from the tractor. Except for its propensity to clog with surface trash,

this tool is easy to work with, and requires little in the way of maintenance or horsepower.

Whether stirring the soil for weed control or for seedbed preparation, it is hard to surpass disk tools for efficiency. The disk blade uses a lifting action in moving the soil (like the moldboard plow), but it has a minimum of compression. In the case of disk harrows, particularly the tandem or offset types, the soil is both loosened and mixed, and, once the tool has gone over the ground, it is left level.

Tandem disk blades are normally 16 to 18 inches in diameter, and are set about 7 inches apart. Mounted in four-gang sets, two front gangs and two

Spring-Tooth Harrows

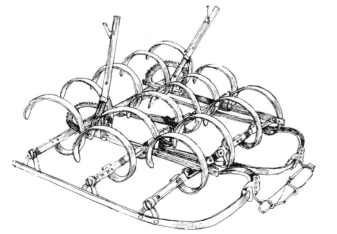

Ganged Disk Harrows

rear gangs, these harrows have different kinds of joining frameworks depending upon the soils to be worked. Rigid frameworks, for example, are used in loamy, non-rocky kinds of soils, while more flexible "floating" frameworks are necessary on stony lands.

Offset disk harrows have larger disks (20- to 24-inch) than the tandem, and they are spaced at 9-inch intervals. Coming in overall widths ranging from 4 to 13 feet, they are arranged in two gangs — one in front of the other. This type of harrow is particularly favored by orchardists, because its offset draft position can be altered to a location outside the rear wheel line of the tractor. This makes it easier to stir soil close to trees.

Both the tandem and offset disk harrows come in three-point hitch models and in trailer types. The trailer-type tandem disks require weighting of the rear sections for even cutting effect, and the trailed offset unit needs adjustable transport wheels for depth control. Both rely for penetration on the angled opposition of the gangs of disks, and upon the cup of each individual disk.

Of all the soil-stirring tools the disk harrow is the most versatile. It is one of the most valued tools in the small-scale farmer's toolshed.

PLOWS AND ROTARY TILLERS

Breaking soil with the plow or rotary tiller utilizes more horsepower hours than any other of the practices in crop production. The purpose of both kinds of tools is to loosen and pulverize the soil, and, when applicable, to mix in the surface trash.

Rotary tillers accomplish this function with L-shaped rotating blades that are mounted on a common, horizontal shaft. They rely on impact for their pulverizing function, and therefore move faster and are subject to more frequent contact with the soil than stationary tilling tools. As the blades rotate they cut into the soil as the tool is pulled forward. This results in a forward thrust that is affected by the pressure applied to force the blades deeper into the soil (static weight in trailer units), and by the braking action of the drive wheels of the tractor.

These tillers are most commonly used with independent engines on smaller pieces of land. Their justifiable popularity springs from their small-sized handiness and the fact that a once-over pass leaves a ready-to-plant seedbed. Lack of engineering is the only reason that this type of tool has not been adapted for more common use on larger acreages. The present state of larger rotary tiller hardware is that it requires considerable horsepower — more perhaps than the small-scale farmer is willing to invest in — and it suffers excessive wear on blades, drive parts, and bearings. It is available in tractor-mount units and in trailer types.

By contrast, the moldboard plow is a time-proven tool for use on large acreages, and it is manageable in one-, two- or three-plow units by most of the tractors in the 20- to 40-horsepower range. The shape of the moldboard plow affects all phases of its functional efficiency. Years of trial have shown that certain shapes are better suited to specific kinds of

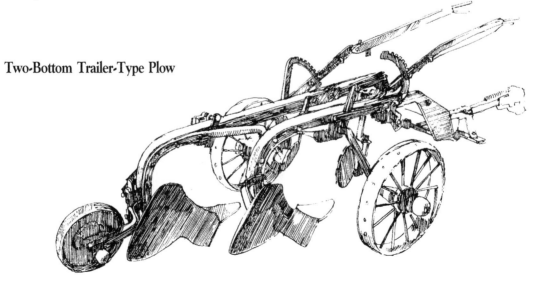

Two-Bottom Trailer-Type Plow

How a Plow Works

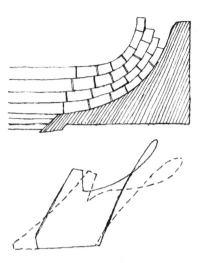

Soil is broken and pulverized as it rides up the moldboard. A 35° angle works best at high speeds, a 45° angle is more efficient at low speeds.

soil and drafting circumstances than others. A stubble-bottom plow, for example, will give maximum pulverization of the soil, but will not scour as well as the blackland type, and the all-purpose bottom is a compromise for several circumstances. As one farmer put it, "My all-purpose plow bottom doesn't do any plowing chore well — but then it doesn't do any of them badly, either."

In ascertaining which bottom is best suited to his individual needs, the farmer should understand the three functional elements of all moldboard plows:

1. The *share*, or nose of the bottom, is the wedge for breaking the soil, and its shape is partially responsible for the bottom's penetration characteristics (the height of the draft point is the other primary consideration).

2. The middle area of the bottom is the pulverization zone that determines the frequency of shear planes in the plowed soil.

3. The turning, or inversion, "wing" area of the bottom is responsible for turning the furrow over. Complete inversion is essential where follow-up cultivation is likely to occur — as with row crops — and many bottoms also employ cover boards, jointers, or trash wires to supplement the function of this section of the bottom.

Positioning the bottom for maximum effectiveness is the product of coulter placement, beam alignment, and proper hitching. If, for example, the trailer plow is hitched too low, the share will not penetrate properly, or if the draft point is too far to the right or left, the furrows will be uneven and the tractor will labor. Bent beams also result in uneven furrowing and they sometimes induce the plow to "walk out" of its proper line. To prevent bent beams or damage to the plow bottom, most plows are equipped with tension-type trips which will release the individual bottoms or, as is the case with a trailer unit, the whole plow.

Farmers working small acreages will find that both types (tractor-mounted and trailer types) have their advantages and disadvantages. The tractor-mounted plow requires considerable time to mount and dismount, but this is offset by its compact handiness when in the field. Trailer plows need only to be hitched up to go to work, but because of their lightness, they tend to "walk out" of the furrow on sidehill plowing.

CHAPTER 4

Farm Practices

SEEDBED PREPARATION

Methods of seedbed preparation vary depending upon the crop to be planted, the amount of trash to be incorporated into the soil, the type of soil to be worked, and the kind(s) of tools the farmer has to work with. Many farmers who, for example, are following corn with a crop like oats or green manure crops like rye or buckwheat, simply crosshatch the surface of the acreage with a disk harrow before applying the new seed.

PLOWING

But with sod or a heavily trashed surface, the farmer usually spreads his fertilizer(s) or soil conditioner(s) prior to plowing or rotary tilling. The latter practice is customarily a two-phase operation: once over the acreage with the rotary hood up, and the rotary blades set just deep enough to cut below the shallow-root level; and then a second run a week or so later at normal plowing depth. The reasoning behind this two-phase practice is that the first run is a weed control measure that throws the uprooted grasses and weeds to the surface where they die before being turned into the soil on the second run. After the second pass, the soil is ready for seeding — unless there is stoning to be done.

Plowing has, for centuries, been done with the moldboard plow. In his book, *Plowman's Folly*, author Edward H. Faulkner contends that farmers habitually plow the soil simply "because they like to plow," and that plowing — moldboard plowing in particular — is unnecessary and even bad for the soil. When written (in the early 1940s), Faulkner's book was a startling and thought-provoking series of what then could be termed "radical" ideas. A great deal of what he had to say made sense, and he was obviously a pragmatic farmer who presented his observations in an appealingly straightforward manner. It is ironic that much of his "heretical" thought is now everyday policy of the same recalcitrant official governmental agencies that he spent his lifetime trying to persuade.

Faulkner was right, in a general sense, when he

Opening and Plowing a Field

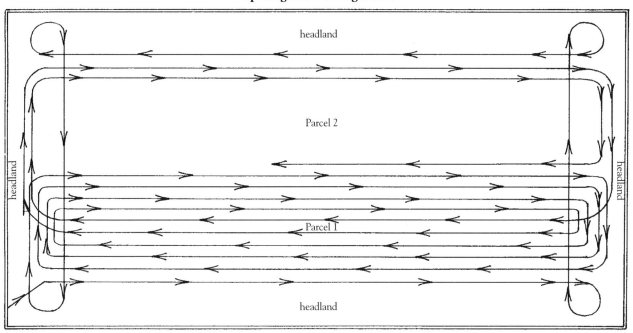

1. *Stake out the headlands, a space in which to maneuver usually twice the length of tractor and plow.*
2. *Plow around the field as staked.*
3. *Continue plowing in Parcel 1 as shown until there is no longer turn-around space. Move on to Parcel 2, turning around in the final furrows of Parcel 1.*
4. *When the parcels are finished, begin at one corner and plow the untouched headlands around the field.*

contended that it was wasteful of soils to stir them. Intensive cultivation and constantly having land open does guarantee erosion. This admonition is particularly appropriate for regions having warm, humid climates. Here the destruction of the soil's organic material by oxidation can be ruinous.

But Faulkner wore zealot's blinders. He was a man of but one view, and, as Disraeli warned, "Beware of the man of one book." The moldboard plow is needed to incorporate organic materials into the soil, and at no time in our agricultural history have we had a more urgent need for this particular practice. Plowing is also essential when opening or clearing new land, or when a stony condition prevents the growth of a desired crop. Lastly, the farm practices of the past thirty years have resulted in too-frequent cases of cemented soils that defy any loosening or tilling instrument other than the plow.

The first question that nearly all new farmers confront in seedbed preparation is when and how deep to plow. Is fall or spring plowing better? Should one plow to the subsoil layer, or into the subsoil?

These questions have no easy answers. Farmers in one locality may regularly spring-plow at 3 or 4 inches with good results, and farmers in another may fall-plow at 12 inches with equally good results. Obviously the answers turn around the kinds of soil the farmer is working with, the fertilizer (if any) he is applying, the crop he intends to seed, and the local climate.

But as a general rule, farmers have found that fall plowing is better than spring plowing. Winter gives the land time to settle and reestablish its capillary connections, and the freezing and thawing cycles tend to break up clods and clumps. If left rough after plowing, the soil stands a better chance of holding the spring's runoff on the land, and this curtails erosion and aids in storing up needed soil moisture. Though the extra soil moisture may not be really sought after by the farmer having heavy clay soil, he will also want to fall-plow his land to avoid getting onto his land late in the spring.

Having stated the general rule, it is now appropriate to make exceptions to it. Land that suffers severe infestations of weeds is best plowed in the

Spring Plowing   **Fall Plowing**

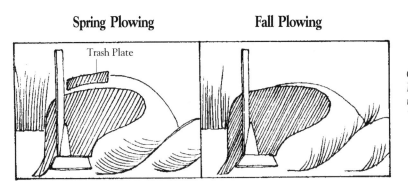

Turning trash completely under with a bolted-on trash plate makes seedbed preparation easier.

Trash Plate

Open soil invites erosion. Plow in the fall without trash plates and at slower "slab furrowing" speeds.

spring. Turning weeds under in the fall simply tends to preserve and protect the seed from the exposure to winter elements. Spring plowing is also appropriate for regions where winter winds are apt to erode unfrozen, open soil

The depth to which the farmer plows depends in part upon the type of topsoil, and its depth. Clay soils, for example, are usually more fertile than sandy ones, and their food elements are made available by a deeper-running plowshare. But if the topsoils are thin, it makes no sense to plow beyond them, unless the farmer is deliberately trying to increase the depth of his topsoil layer by mixing in subsoil. This method of "soil building" should be done cautiously and slowly. An inch lowering of plow depth per plowing is plenty. Each plowing should also incorporate plenty of green organic and fertilizer materials.

Other factors, like the kind of crop to be planted, may well affect the plowing depth. Land for small grain crops, for example, is frequently plowed at shallow depths (say 3 or 4 inches) on good soils, while root crops may require loose, friable soil to a 10-inch depth.

The newcomer to farming is wise to find out the typical plowing practices employed in his area for the same kinds of crops that he plans to grow, and, unless he finds evidence to the contrary, he should mimic them. In the absence of any local guidelines he should fall-plow at a 6- to 8-inch depth, and, when possible, he should plow, seed, and grow sample plots to test pragmatically varying practices. "Show me" is not a philosophy that should be restricted to childhood education.

There is probably no more exciting moment in the new farmer's experience than when he opens his first field, providing he knows how to go about it. Over the years farmers have developed several ways to plow a field, but the most common when using a

two (or more) bottom outfit, is the "parcel method."

To start, stake the *headlands* (the area at each end of the field that is used to turn around), leaving about two times the length of the tractor and plow, and then plow all the way around the field between the stakes (turn the furrows in toward the center). Use the furrows as a guide for when to raise and lower the plow at the beginning or end of each run. Begin plowing the center by dividing the field into parcels and plowing furrows that progress inward toward the center of the first parcel until the U-turn in the headland is too tight (see illustration, page 37). At this point, make long U-turns to begin plowing the second parcel, and continue until all parcels are plowed. To finish the field, start at an interior corner of the headlands, and plow round-and-round (turning the furrows in) until the headlands are also done.

Experienced farmers take a great deal of pride in their plowing, and the new farmer would do well to emulate them. Straight, evenly turned furrows, except when contouring, make for even coverage, and easier work in the following refining processes. When finishing a field or a parcel, one should take the extra time to fill the dead furrow by plowing shallow furrows back into the space. These extra runs will insure that the dead furrow will not become an erosion channel, and it will save the farmer's kidneys from unwelcome shocks when working against the grain of the plowing in any of the subsequent field operations.

No discussion of plowing practices would be complete without mention of the tiller disk plow and the lister-bedders. The former is popular in the grain-growing areas of the Midwest and in regions where the soil is heavy and free of stones. Although the disk plow does not bury trash as well as the moldboard, its time- and energy-saving advantages make it an ideal tool for specific areas. The lister plow was, and still is, used in the South and the

Spike-Tooth Smoothing Harrow

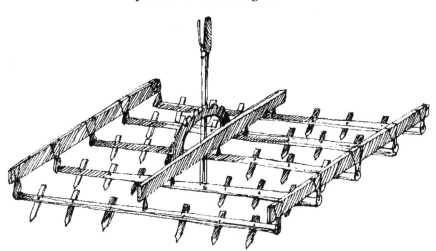

Southwest. It is a "middlebusting" type plow that throws soil in both directions, thus creating deep ridges and furrows. For irrigated bedding crops it is a desirable, multipurpose tool.

HARROWING

After using the moldboard plow, the soil is left ridged and rough. It requires refinement before seeding and planting, and this is the primary purpose of the disk harrow. Most farmers use the tandem type, but the offset type is growing in popularity. Light soils can usually be leveled and pulverized with a once-over pass of the tandem disk harrow, but some heavier soils or soils that require stoning may require a second lapping or crisscrossing coverage. If stoning is necessary, it is usually best to make a single pass over the field to "discover" the stones, and then to follow the stone-picking with a smoothing pass.

For many crops, even in the heaviest of soils, disk harrowing after plowing provides an adequate seedbed, but other crops, especially fine-seeded ones, may require the use of a smoothing (peg, spike-toothed) harrow before seeding. This harrow is commonly ganged, two or three sleds to a gang, and dragged once over the field. Each sled has five rows of diamond-shaped pegs (reversible and adjustable) that afford surface penetration and further pulverization of the seedbed. With the teeth set shallowly, the smoothing harrow is also occasionally used to bury broadcasted seed. This practice has the effect of drilling, as the seed falls into the small, evenly-spaced furrows made by the spaced pegs. Covering then occurs as the loose soil falls back into the furrow following the passage of the peg.

Smoothing or spring-toothed harrows are also used to kill newly germinated weeds. Often, rainy weather will interrupt seedbed preparation, and one or two weeks may pass before the farmer can get back to the job. Shallowly set teeth on either harrow will loosen the rain-compacted soil, and uproot the growth of young weeds.

SEEDING AND PLANTING

Seeding and planting should follow seedbed preparation as quickly as possible in order to take advantage of the loose, friable soil medium — not to mention the farmer's personal urgency to get his crop in the ground. Frequently a farmer will find himself racing against the weather. He knows the benefits that accrue from a couple of days of gentle rain following seeding, and if rain is promised, he may not give the attention to seeding that he should. He must, however, remind himself of the importance of adequate and even coverage. A break in broadcast seed coverage provides fertile ground for the growth of weeds, and will complicate the harvesting process later on. Uneven coverage with row crops that will need later cultivation and harvesting with spaced row-type equipment will be an equally exasperating headache.

Broadcast seeding is probably the best technique for most new small-scale farmers who want to seed grasses or many of the grains. The cheapest, most efficient tool for this purpose is the hand-cranked, spinner-type spreader. Depending upon the size of the seed, these small hand units have a lateral spread width of 15 to 25 feet. Assuming a steady walking speed, one can seed about 2 acres per hour (counting refill and marker-moving times).

Seeding large fields with this hand-cranked seeder requires the use of flagged indicator staves to mark the seeding pattern. A good way to use these staves is to test the spreading width of the seeder — watch for the outer puffs caused by the seed striking the loose soil — and then to pace off this spread distance. Divide this paced distance in half for the first row, and place one marker at the beginning and one at the end of the spreading row. Then pace off the full spreading width and place a second set of markers for the next row. At the conclusion of seeding the first row, leapfrog the first set of markers to the third row, and so on until the field is done. To maintain a straight line between markers, the trick is to line up a background object with the distant marker at the beginning, and to periodically check on that alignment. Undulating ground or very long seeding rows may require extra mid-row markers.

Machine-powered seeders and planters are essential tools for large field plantings of corn, beans, potatoes, cotton, peanuts, and most vegetables. While the bulk of these machines are specialized for a particular crop, the typical corn planter and the seed drill offer a versatility that should appeal to the beginning small-scale farmer. With the appropriate seed-cell plates, a corn planter will also plant a variety of beans, and even peas. Seed drills will sow most grasses, clover, alfalfa, oats, or wheat. Attachments for them, like the wheel press, eliminate the need for follow-up land-rolling. Drills are particularly favored on stony land, land that has recently been cleared and is still full of roots, and on land where the farmer desires to intertill a broadcast green manure crop between the drills.

On loose, loamy soils, farmers sometimes skip the seedbed preparation described above. Many farmers, for example, merely broadcast, drill, or plant over unplowed, undisked land if it is of good tilth. With broadcast seed, the only tillage this kind of land sees is that of the disk harrow, and this once-over treatment is only employed to cover the seed.

CULTIVATING

When weed killers have not been used with row crops, planting is followed by cultivation. Corn and beans, for example, are usually given at least two cultivations, and, occasionally, three. Most cultivators are edged with sweeps or shovels to weed and loosen the soil between the rows, and with disk hillers to throw smothering dirt over weeds that are growing in the crop row.

Operating a cultivator on row crops requires a great deal of precision because of the proximity one must maintain to the rowed crop. When using a two- (or more) row cultivator, the evenness of row spacing is critical. Poorly spaced rows will be gouged out by the cultivator and later by the harvesting equipment.

The new small-scale farmer will soon learn that in the area of farm practices there are no absolutes. One can, for example, use cultivators instead of herbicides, but this practice also has its price. The breakdown and oxidation of surface soil particles that occur with clean cultivation is a problem to be reckoned with. With the breakdown of the soils comes quick rainfall sealing of the soils, with consequent reduced aeration and increased erosion. Further, the compaction of the layer below the cutting depth of the cultivator's edges reduces the root-level porosity of the soil. As a general rule, farmers in the South and in parts of the Wheat Belt where surface oxidation is more severe have come to recognize this problem and now successfully employ some of Edward Faulkner's "trash farming" techniques. These practices entail keeping their soils covered with a surface mulch of crop residues, and stirring their soils as little as possible. Farmers in the Corn Belt and the Northeast, however, have found the conventional tillage practices still appropriate and sustaining of good yields.

HARVESTING

Good yields are, after all, an integral part of good farming practices, and harvest time is when the farmer reaps what he has sown. There are machines to harvest virtually every crop grown, but many of them are so specialized (and expensive) that the small-scale farmer should think twice before planting a crop that requires their use.

It is, of course, important to know how to harvest a particular crop, but for the most part, the use of harvesting equipment is self-explanatory. The most basic and important knowledge a beginning farmer needs to acquire is when, rather than how, to harvest.

Hay, for example, is a crop that has for years suffered from harvesting ignorance or neglect. Look over any hay dealer's stockpile and you will invariably find that most of it was cut after the grass had passed its prime. It is usually stemmy and rank, showing only a small proportion in leaf — and it is the leaf of the hay that is nutritious, not the stem. Every kind of hay has its own signal of when it should be cut, and no two are alike. For example, June grass should show a fully developed seed head before cutting, but the head should still be green. Timothy, on the other hand, should be cut before the seed head appears, or just as it is emerging from its sheath. Alfalfa is prime just before it blossoms, but clover and trefoil should not be cut until their blossoms have reached full maturity — an ideal circumstance for beekeepers.

Second- or third-cut rowen (when hand-mowing, the first cut was known as a "math," and subsequent cuts as "aftermaths") are frequently determined by the length of season, rather than the maturity of the grass. In areas that have short growing seasons, many farmers find the second cut too short for baling, and they either chop it for haylage, or mow it for sheet composting.

Wheat and oats should be harvested only when most of the stems have lost their greenness, the grain head yellowed with the grain past the soft-doughy stage, and when the grain can be threshed out by rubbing the heads between the palms of the hands. But, like dried beans, if the grains are allowed to dry too long in the field there may be unrecoverable losses due to lodging, or to premature threshing while mowing. As a general rule, beans should be cut when the pods are well-formed (the lower ones filled), and when the leaves of the plants have begun to yellow. In a pinch — say a rainy harvest season — these crops may be cut and stored under cover (ventilated stacks) to be threshed after they have dried.

Corn offers considerably more latitude in its harvesting. Rainy weather, except as it affects getting onto the fields with equipment, does not materially affect the harvesting practice of a small-scale operation. Sweet corn can be readily hand-picked during rainy weather, and the stover or fodder left for a drier time. Field corn intended for ensilage is the most common of the corn enterprises, and it can be harvested (chopped and shredded) rain or shine. Most farmers chop their corn when the silk of the ear has turned dark brown, and when the kernels begin to show dents on their flat upper sections.

By contrast, potatoes are far more demanding of good harvesting weather. They need to be turned up to dry (for an hour or two) on top of the soil, and muddy conditions make for rougher handling and considerable processing losses (bruising). Sweet potatoes should be dug when of size and when test ones are cut open to show that the cut surfaces dry quickly. Those that are unripe will remain moist. Regular Irish potatoes can be harvested to market demands. Early market potatoes are often harvested before they have had time to mature, while the mature, keeping potato is not generally harvested until the vines have died back or have been killed by frost.

The maturation characteristics of these major crops may be affected by regional differences of climate or soil, the technical capabilities of the machinery the farmer has to work with, and the intended use of the crop. These are variables that each newcomer to farming will have to weigh for himself.

The new farmer should research his neighbors' practices before putting a seed in the ground. Should he find that his neighbors know nothing about the crop, he should not necessarily abandon the idea, but he should suspect that there may be a good reason why it is not known in his neighborhood. If he is still determined to plant this particular crop, he should redouble his research efforts.

He might find, for example, that his farm is in an area that annually suffers rains during the normal hay-gathering season, and that the time-consuming, old practices of mowing, raking, tedding, and baling resulted in too many instances of ruined hay. He might then logically consider machinery that would hurry up the haymaking process. Or he may discover that his heavy soils and short growing season make a necessary early planting of wheat a risky affair, and an early harvest with subsequent heat processing of the seed a necessity.

PROCESSING AND STORAGE

Some crops are processed in the field, and are occasionally sold there. This practice saves the farmer both money and time. He does not have to bear the costs of storage, transport, or the equipment involved in further processing. Hay, for example, is frequently sold "behind the baler," and occasionally the farmer will sell his grains "off the combine."

But most crops require further processing and storage, particularly when the farmer plans to use them for his own livestock. Hay that the farmer plans to store (for later resale or for his own use) must be properly cured and dried before being stacked under cover. Weather does not always cooperate with the harvesting process, and hay that has gotten wet by rain, or hay that was baled with morning or evening dampness is a potential source of mildew, mold, or spontaneous combustion. Artificial heat provided by fan-blown hay driers can correct this situation, or the farmer can loose-stack the hay under cover so that there is plenty of air circulation. After it has dried he can stack it for storage.

The problem in making properly cured hay is to dry the stems before the leaves become so dry and brittle that they are lost in the handling. In warm, humid, or rainy regions farmers use a hay crusher (conditioner) after mowing to break and crush these stems and thus hasten drying. Machines like the hay crusher or the haybine (a machine that combines mowing, crushing, and swathing) can reduce the curing time by as much as 30 to 50 percent. But even in the best of conditions, some loss of leaves cannot be avoided with field curing.

Similarly, there are some unavoidable field losses in combining grains like wheat, buckwheat, barley, or oats. Drilled grain tends to mature more evenly, but with any seeding there are bound to be seed heads so dry that they shatter upon first impact with the reel or the reciprocating knife of the combine. If the harvesting is very late, losses can go as high as 10 to 12 bushels per acre. To reduce these losses, many farmers harvest early, then use heat augers to dry the grain before storing. The exception to this early harvesting practice is barley: tests have shown that barley yields better quality and quantity when harvested after the stems have turned completely yellow.

If there was little weed growth in the field, it is possible that the combine has left the grain sufficiently clean, but in most instances it must be run through a fanning (cleaning) mill before drying. This cleaning will eliminate any moisture-holding chaff or weed seeds. It can then be dried with heat augers set into the storage bin, or if there is no bin available, the grain can be piled in shallow layers (preferably on screens) and turned over frequently with a shovel. Once dried, it can be stored (with guards against rodents), or it can be subjected to further processing (hammer mills, dehullers, rollers, etc.), depending upon the use intended.

Dried beans are much like grains in their processing. Assuming that the beans are dry after combining, they are cleaned in a fanning mill and then stored. Like the grains, they are also prone to damage by heating or molding, and they are not safely stored if the moisture content exceeds 14 percent. Obviously a moisture meter is a useful, if not essential, tool for growers of grains or beans.

Ensilage corn is a desirable protein feed for livestock for several reasons. Easy processing is one of them. After chopping in the field, it is transported to and loaded into the silo. There are two kinds of silos presently in use: the vertical silo (made of wood, concrete, or fiberglass), and the horizontal silo (made of wood or concrete or, in a pinch, earthen-sided). The expense of vertical silos has made the trench or bunk silo a popular alternative for dairying or cattle-raising farmers. This cheaply constructed, horizontal silo can be built above or below the ground, with wood (portable bunks), or with more permanent concrete sides. Plans for these silos are available from most local agricultural agents, or from the USDA.

Silos are intended for storage of four or five months, and they depend for their success upon a pickling fermentation action brought about by the same kinds of anaerobic bacteria that produce acid and alcohol. Nutritious, long-lasting silage (both corn and hay) is made by:

1. A finely-chopped crop.
2. A tight packing to eliminate trapped oxygen (usually accomplished by driving a tractor back and forth over the silage).
3. An airtight seal of plastic or other impermeable layer.

Ideal packing moisture content of the silage is between 65 and 70 percent, and a pH of 3.5 to 4.5 is the desirable reading for properly pickled silage.

Of the crops discussed above, potatoes are the most easily damaged during harvest and transport, and they are the most finicky about storage conditions. It is because of the difficulty of storage that most farmers sell directly to warehousers or potato dealers. The small-scale farmer, however, may be trying to sell his crop directly to consumers, and he will therefore have to make provision for storing his crop for several weeks, or even for two or three months.

Detailed plans for constructing a potato storage building are obtainable from the USDA or nearby agricultural experiment stations, but most small farmers can make do with a cellar, or with pit storage. Temperatures of the storage area should be high enough to keep the potatoes from freezing, but low enough to keep the tubers dormant (35° to 40°F. — 2° to 5°C.). Light must be excluded, and the inverted, V-shaped piles should not exceed 6 feet in depth. If a pit is used, it should be located in a well-drained spot, and ventilation should be provided by building the piles around wooden or plastic (pipe) flues. These flues should extend from near the bottom of the pit (trigged up with stones) to well above the covering. During freezing weather these ventilators should be stuffed with insulation and topped.

During the first few weeks of pit storage, the crop can be topped with dry straw, and just enough soil to exclude light and freezing temperatures. But the depth of the straw and soil should increase (in increments of 8-inch layers) as the outside temperatures fall.

IRRIGATION

Questions of ownership of water have plagued man since he first began to farm, and because irrigation is a necessary farm practice in many parts of the country, these questions remain subjects of heated debate today.

In the United States there are two fundamental systems of water rights, riparian and prior appropriation. At the risk of oversimplification, riparian rights are those claimed by the upstream owner of land through or by which a stream may flow. He claims the right to use this water without regard to when he may use it, or whether he uses it at all. Prior appropriation rights give the first *user* of water from a stream (though he may not own land on the stream) a given amount of water, at a given time and place, and these rights remain with him

and his heirs so long as the use to which he puts the water is "beneficial."

It is not within the scope of this book to give the water access laws for the country, but as a general rule, the riparian water doctrine is still (with modifications) recognized in the eastern states, and has varying credibility in California, Washington, Oregon, Texas, Nebraska, Kansas, North Dakota, and South Dakota. It has been specifically denied as a doctrine in the Southwest (Nevada, Utah, Colorado, New Mexico, Arizona), and in Idaho, Montana, and Wyoming. Where the riparian doctrine is denied or in question, the doctrine of prior appropriation has taken its place.

Obviously water law is legal and political dynamite, and a new farmer settling on farmland where irrigation is a necessity should come to know his local water law as well as he knows his tractor or his spouse (whichever he holds to be more important).

But knowledge of the law is of no help if the water quality is below that needed for irrigation. It may be overloaded with silt, brush, or other debris, or, as is more commonly the case, it may be saturated with dissolved salts. To help solve these problems, the new farmer should enlist the help of the Soil Conservation Service. SCS agents will aid the farmer in determining the extent of the problem, how it affects his particular soils, and how the water quality would affect the crops he might want to grow.

As a practice, irrigation was once confined to row crops in the West and Southwest, but since the 1940s the practice has extended into the South, parts of the Midwest, and spottily into regions along the East Coast. This phenomenon is due to increasing crop specialization and economic pressures to increase and standardize crop yields. Furrow and flood irrigation are still the most commonly used methods, but sprinkler systems have become increasingly popular, despite their high costs and engineering complexity.

Sprinkler systems are still most often used in "borderline" farming situations — that is, where the natural moisture supply is erratic or insufficient for maximum and consistent crop yields. Before committing himself to an expensive supplementary sprinkler-type irrigation system, the new farmer should exhaust the more conventional practices of keeping what natural moisture is available to him back on the land, and work at conditioning his soil's receptivity to and utilization of water.

But no amount of soil conditioning will replace requisite quantities of water, and in arid farm areas furrow or flood irrigation is a necessity. Most farmlands in these areas are already crosshatched with a regulated regional water supply system, and it remains only for the newcomer to join or "buy into" his local water conservation or irrigation district.

Once he has tapped into the local irrigation water supply, the costs and methods employed for use of the water are the farmer's own. Again, help from the Soil Conservation Service is indispensable. SCS agents will help the farmer to determine what land is feasible to irrigate (soil type and slope) and when and how much water he will need (specific crops and soil condition); they will also suggest the most economical and efficient ways to deliver the water over his fields.

All on-the-farm irrigation begins with the supply canal or ditch. Water levels should be a minimum of one foot above the highest point of the field, and the canal or ditch should have the necessary headgate division boxes, turnouts, or siphons to divert the water to the appropriate field(s). Furrow irrigation depends upon gravity, and it is used for nearly all cultivated crops. Water is drawn or diverted from a head ditch that runs across the upper edge of the field to the furrows. The furrows run down or slantwise across the slope of the land.

For best use of the water this slope should not exceed 0.25 percent for row crops or 5 to 6 percent for cover crops. Length and depth of furrow are determined by the potential rate of water flow (from artificial or natural sources), and the absorptive rate of a particular field's soil. These factors involve potentially dangerous erosion conditions, and they should be determined only after consultation with a Soil Conservation Service agent.

Flood irrigation is less often used, but has distinct advantages with many close-growing crops and with orchards. This method requires relatively flat land and larger streams of water than are provided for furrow irrigation. Parallel flood strips, usually 50 to 100 feet wide, are created by plowing up bordering terraces to confine the waters. The object of this system is to flood the confined area as rapidly as possible to a depth determined by the measured intake rate of the field's soils, and then to allow the water to soak in. Rate of flooding and intake rates should also be arrived at with the aid of the local SCS agent.

Contour Plowing

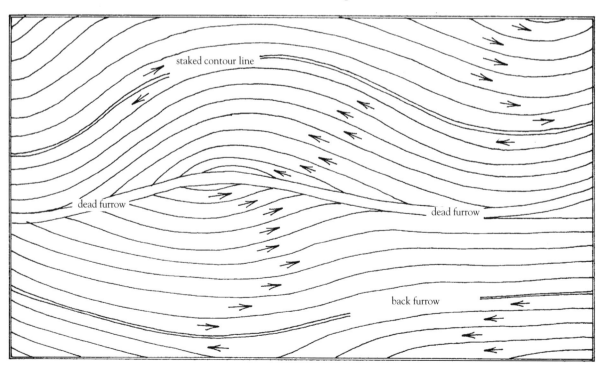

OTHER SOIL CONSERVATION PRACTICES

As with access to water, there are special laws to handle the drainage of it, and the new farmer who anticipates a need to alter the natural drainage on his farm should first determine the local law relating to the disposition of surface waters. Drainage law is complex, and varies from state to state. In most of them, this complexity has led to the formation of drainage enterprises (often called districts), which concern themselves with whole watersheds. In fact, about one-fourth of all cultivated fields in this country already come under the aegis of an existing drainage enterprise.

If drainage has made one acre of tillage available to the farmer that he did not have before, conservation practices aimed at keeping water on the land have given him two. The plain fact is that more of our farmland suffers from a lack of water than from an overabundance of it. Two fundamental practices, terracing and contouring, are employed across the country to help control the loss of surface water and the consequent loss of topsoil.

Terracing is a method of breaking up a long, water-wasting slope in a field. Cut across the slope of the land, a terrace is an artificial "shelf" that can serve as an absorptive pause in a runoff pattern, or it can be a planned waterway with a ridge on the downhill side that is designed to dispose of an otherwise erosive flood of water. In most instances, the small-scale farmer will find that he can move enough earth with his plow to make any terracing that he will need, and he will not have to resort to hiring outside earth-moving equipment.

Channel terracing should be surveyed, staked, and uniformly graded at 0.4 to 0.5 percent, so that the collected water flows to a desired outlet point. Care should be taken to keep farm roads from crossing the line of the terracing, and, in general, all tilling operations should be done parallel to the terrace line. If more than one terrace is required on a slope, it should be measured with a tape so that it runs parallel to the existing one.

Contour farming is, in a sense, miniature terrace farming. Employing this practice, the farmer plants across the slope, and sometimes alternates strips of grass crops with clean-tilled crops like corn or beans (strip cropping). With the plowing and planting done on a contour line, the small ridges and (later) the crop stems act as deterrents to surface runoff and soil erosion.

Field topography seldom allows evenly spaced contour lines, and it is the usual practice to plow or seed the upper and lower boundaries of a strip first. This means that some of the in-between plowing and planting furrows or rows will be incomplete, as the plow or planter has to be lifted before the end of the field is reached. These are called *point rows*. When possible, it is a good plowing practice to leave the dead furrow midway between the upper and lower contoured back furrows (see drawing, page 44).

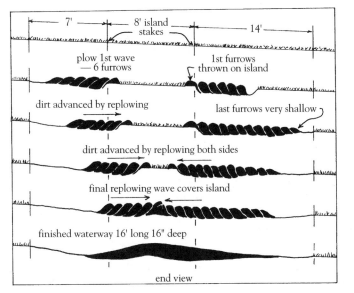

7'   8' island stakes   14'

plow 1st wave — 6 furrows

1st furrows thrown on island

dirt advanced by replowing

last furrows very shallow

dirt advanced by replowing both sides

final replowing wave covers island

finished waterway 16' long 16" deep

end view

Making a Terraced Waterway

The secret to terracing is to begin plowing with a shallow penetration of around 4 inches. Then with each replowing, set the plow lower so the points find solid purchase in undisturbed ground.

CHAPTER 5

Cash Crops

Growing a profitable cash crop today on the small-scale, polycultural farm involves careful planning and a hard-headed approach to details. Looking at the following lists of dos and don'ts, one can detect the old-fashioned horse sense that is the legacy of a hundred years of small-farm experience.

Dos

1. Do grow labor-intensive crops — or, to put it another way, do grow crops in a labor-intensive fashion.
2. Do diversify. Grow two or more cash crops as a hedge against crop failure.
3. Do experiment. Flexibility and a willingness to try new methods are indispensable qualities for any farmer.
4. Find and/or create market "niches" (see below), and fill them completely.

Don'ts

1. Do not market to middlemen. Cultivate a farmer-to-consumer market.
2. Do not hire outside labor if you can help it. Restrict the scale of the venture to what you can cope with by yourself.
3. Do not acquire interest-debt burdens.

Like any other enterprise, successful small-scale farming is the product of maximizing profits and minimizing expenses, and the first maxim of maximizing profits is for the farmer to keep as much of them for himself as he can. Interest payments and outside labor costs make everyone rich except the farmer. Because of this he should give serious study to the scale of his operation and to the kind(s) of cash crops beforehand to avoid these pitfalls.

Among the other pitfalls the new small-scale

farmer faces, that of marketing is perhaps the most serious. The first of the don'ts (do not market to middlemen) is merely an extension of the logic of keeping one's profits at home. Too great a proportion of today's retail food prices are taken by middlemen, and if the small farmer is alert and imaginative he can claim this profit for himself.

The small-scale farmer should not deceive himself into thinking that direct marketing to the consumer is merely an option to be exercised before reverting to wholesale marketing. The unvarnished truth is that as long as agribusiness is subsidized by tax dollars, the small-scale farmer will have a hard, if not impossible, time competing on the wholesale market. Pregrowing contracts with retail co-ops, like farmers' markets, have many attractive features, the foremost one being that the farmer sells directly to the consumer and keeps all of the profit for himself (excepting, perhaps, space rental or license fees). The disadvantages of farmers' markets are few, but they are significant. The farmer must transport his crop (in some cases this can be a significant factor) to the marketplace, and he must invest whatever time is necessary at the marketplace to sell his crop. Unless the market is open on a daily basis, the farmer who has a crop that requires daily picking must assume the additional expense of storing perishables.

The ideal circumstance is for the farmer to sell directly to the consumer *from his own farm*. This invariably involves research and, sometimes, some creative finagling. The first time we ever grew oats into our rotation, for example, we could not find anyone who would pay a decent price for the oats as they came from the threshing combine. Everyone wanted them hulled, rolled, and delivered, which meant that we had to either purchase a lot of new equipment or go through an expensive and time-consuming processing step.

While we were thrashing the problem around, a houseguest who happened to be an archer humbly suggested that we might call some archery target makers, to see if they might be interested in the oats by-product, straw. To make a long story short, we did, they were, and we ended up virtually giving away the threshed, unhulled, unrolled oats, while selling the straw for four times more than we would have made from the grain alone. To top it all off, the target-making firm was so delighted with us as a source that they agreed to pick up the baled straw at our barn!

Enticing the consumer to the farm, even from relatively distant points, is often the easiest part of the marketing problem to solve. City people, for example, are very positive about visiting a working farm. Our neighbors recognized this early on when they were selling hay. Little research was needed for them to find that suburbanites who kept only a few head of livestock were willing and able to pay more than double what local Vermonters offered for their hay.

The problem, as our neighbors saw it, was to find a way to get these suburbanites to come haul their own feed, *and to pay them* (our neighbors) *for the privilege*. They took out an ad in a suburban weekly, which priced their hay at the going rate in that particular suburb, and they took advantage of their large, old farmhouse by offering the following sweetening:

> . . . *or you can haul your own from our farm (minimum 2 tons) for half price. For that price, we'll help you load, and then we'll throw in overnight lodging and a great breakfast for two at our farm when you come to get it . . . .*

Two weeks later they had sold their entire summer's cuttings, and those customers have already spoken for next year's crop in advance. This is what I call creating a market "niche" and then filling it. Another ideal niche, of course, is to be able to sell your product from a roadside stand, but not everyone has land adjacent to or near a well-traveled road or highway. Since many newcomers to small-scale farming are located on remote back-country roads, direct consumer marketing can pose a significant problem.

The answer to this problem will vary with proximity to large population centers and the type(s) of crop grown. *But if the price and quality of the crop are appealing, the consumer will come to the farmer.* Established growers of specialty crops for years have counted on this kind of consumer predictability for their livelihood. A reputation of this sort takes time and patience to establish, but is, in the long run, invaluable. A farmer can build up a clientele for his farm-fresh crops, be they eggs, berries, beef, or potatoes, providing he grows a consistently high-quality crop and keeps his prices at a reasonable level. Because he has no transport, storage, or overhead retail costs to contend with, he can usually afford to lower his prices to a competitive level.

Reputations like this take time to establish, so how, the newcomer will rightfully ask, does one guarantee a market for his crop in the interim? This is where alertness, flexibility, and imagination enter the picture. For example, combinations of the various marketing alternatives cited above can be employed. A portion of the farmer's crop (contracted) can be sold to a retail co-op, and another portion could be sold at a local farmers' market.

Another alternative is for the farmer to contract to grow a specific quantity of a crop(s) for individual consumers or groups of consumers. Food conspiracies, an outgrowth of rent strikes on the West Coast, represent a good potential outlet for the small-scale farmer's crops, particularly for staple crops that can be stored. Food conspiracies are nothing more than organized consumers' groups dedicated to bargain-hunting. It stands to reason that the ultimate food bargains can be gotten by dealing directly with the farmer.

If, for example, the farmer were to contact an established food conspiracy, or if he were to set up one of his own by contact with urban individuals (friends, relatives, neighbors, or even strangers attracted by advertising), he could contract to grow a whole year's worth (per family) of storable staples like potatoes, wheat, carrots, squash, onions, beans, etc. This kind of a contract would, of course, be for an agreed-upon quantity, and the farmer should expect and receive an advance, good-faith deposit from his customers.

The advantages of this arrangement to the consumer are considerable. Besides a bargain price, the consumer knows where his food is coming from, and, should he so desire, he can even make arrangements as to how it is grown (e.g., no chemicals). Food contracts offer a different kind of stability and promise to the farmer. He can plan ahead, schedule his crop rotations, make early arrangements for fertilizers, soil amendments, and water, and, while the contracted-for food may not make up his entire crop, he has a set income that he can count on.

Most storable, staple-type crops are harvested at roughly the same time, and if these crops are the ones the farmer is to grow, he can schedule the times that his customers pick up their food to a concentrated time in the fall. One or two weekends will usually suffice to weigh, package, and distribute the crop. Harvest time on the farm is a very special occasion, and the farmer would do well to share the excitement of it with his customers.

Once the farmer has settled on a marketing niche that involves the customer sharing the ambiance of his farm and the harvest, it is imperative that the buyer's expectations be fulfilled — and more! This is another example of filling the niche that one has chosen or created. Customers should be encouraged to smell, taste, feel, and experience the crop(s), the farm, and the special opportunities of the farm's particular locale.

Some nearby personable friends of ours, who make their living from a series of specialized, pick-your-own crops, have deliberately chosen to make their farm a flowering Eden during the harvest season. At first, they merely figured their ornamental flowerbeds into their overhead. Then, as their customers began to demand it, they began selling lush bouquets of their blossoms. This, in turn, has become, over the years, a large and profitable greenhouse operation.

These folks rely heavily on a limited local trade and a choice location. Repeat trade is all-important in their venture, and they recognize the need to sell a little bit of themselves with each customer transaction. Fortunately, they are secure enough in themselves and in their farming skills that they can afford this kind of contact.

Because today's successful small-scale farming relies so much on bringing the customer to the farm, good personal contact skills are a must, and, like one's soils, these skills must be cultivated. If rubbing elbows with customers on a daily basis during the harvest season is not preferred, then the potential farmer should reexamine whether he really wants to get into this kind of endeavor.

But, while I cannot overemphasize the need for personalized marketing in a successful venture, the farmer's primary task is still the day-to-day business of caring for his soils, getting a crop in the ground, and then harvesting it. These rudimentary functions constantly confront him with the need to balance the size of his operation against the amounts of labor required to do each task. Degrees of mechanization will vary with the kinds of crops to be grown, soil types, and climate, but the profit-saving rule that every small-scale farmer needs to apply is to choose the least amount of mechanization he can live with and still get the job done.

Growing grass is the cornerstone of almost all farm cash crops, and knowing how to grow it is fundamental to a successful small-scale farm enterprise. The trick to growing cash-crop grass successfully is advance planning and research. Tight rotations, intertilling, and selecting the right varieties are ingredients that can spell the difference between income years and deficit years. Though grass is often used rotationally to fertilize, protect, and condition the soil, it should always be taxed for its money-earning potential. "Plowing back profits" is a luxury that few farmers, particularly the scratching-to-make-a-living variety, can afford.

The time-honored adage among experienced farmers is, "If you take care of your grasses, your other crops will take care of you." From the representative analyses that follow, the reader should be able to grasp the knack of growing most grasses.

I. ANALYSIS OF SOME GRASS, CLOVER, AND LEGUME CASH CROPS

A. General considerations:

The crops discussed here are: timothy grass, brome grass, medium red clover, and alfalfa.

B. There are three purposes for growing these crops:

1. **Indirect cash crop.** This purpose includes growing grass, clover, and alfalfa for pasture, silage, or for hay that is to be used by the farmer to feed his own livestock.

2. **Direct cash crop.** Hay or pasturage that is grown for direct sale or rent falls into this category. Also included here are seed-growing enterprises.

3. **Soil investment.** Grasses, clovers, or legumes that are included in rotational schedules for green manuring, soil retention, composting, or mulching purposes fit into this category.

C. Rotation schedules:

Except on some annually flooded bottom areas, crops should be regularly rotated. In these rotations clovers or alfalfa usually follow small grains, and grasses usually follow clovers or alfalfa. Newcomers to farming should consult their neighbors and their local SCS or County Agricultural agents to determine rotation schedules that are common to their area. Some typical schedules might look like this:

1st year	2nd year	3rd year	4th year
oats or wheat	red clover	corn	oats or wheat
oats or wheat	alfalfa	alfalfa	timothy
oats or wheat	crimson clover	potatoes	oats or wheat

D. Suitability (by type) of crops:

1. **Timothy.** This is still the most commonly grown perennial hay grass in the country. It is an easily established crop that is suitable (with annual fertilization) for long-term hayfields or pasture. Best suited to areas having cool, moist conditions, it thrives on well-drained loam soils, but it can be grown on clay or silt soils or on acid soils that will not support legumes. Pure timothy is grown only for seed (fall seeding rates: 3–4 lbs. per acre; spring: 8–10 lbs. p.a.) or for horses and mules, and it is most often mixed with clovers and/or alfalfa. Typical mixes are: alfalfa, medium red clover, alsike clover, timothy (4:3:2:3 lbs. p.a., respectively); alfalfa, medium red clover, timothy (6:3:3 lbs. p.a., respectively); medium red clover, alsike clover, timothy (4:2:3 lbs. p.a., respectively). It can be sown in the fall or spring depending upon the mix used and the severity of the winter. As pasturage it should be used rotationally as it will not stand up under continuous, close grazing. Hay should be cut early (while the head is "still in the boot") to provide maximum nutrition. Yields in hay should amount to 2 to 2½ tons per acre. If taken for seed, timothy should be combined while the seed is still in the advanced doughy stage to avoid shattering. Seed yields vary from 3 to 7 bushels per acre (135 to 315 lbs.).

2. **Brome grass.** Excepting orchard grass, brome grass is perhaps the best general perennial pasturage available. It is grown across the northern half of the United States, from the state of Washington to Maine, and it owes its popularity to its dense sodding characteristics and ability to withstand drought. Brome grass thrives on fertile, well-drained clay or silt loam soils, but can be grown on sandy soils if they are well fertilized. In the northern parts of the United States it is usually sown as early in the spring as the ground can be worked, and in southern and central areas it is sown in late summer or early fall. In the latter case it is a good plan to seed brome

grass following a small grain crop. Seedbed preparation should begin as soon after harvesting of the grain as possible. Since a new stand often takes two years to establish itself, it is common practice to plant a thin (reduced by 25 percent) crop of early-maturing grain with the brome. Mixture and rate of seeding depends upon climate, soil, and length of rotation. Short rotations (one to two years) usually employ mixtures of brome grass and clover (8:6 lbs. p.a., respectively); for longer rotations the brome is mixed with alfalfa and medium red clover (8:6:4 lbs. p.a., respectively). In the West irrigation conditions should see 1 lb. p.a. of orchard grass, wheatgrass, or ryegrass added to the above long-rotation mixture. Pure stands of brome grass will yield 300–500 lbs. p.a. in seed when combined.

3. Medium red clover. The most important of the clovers, medium red mixes well with grasses, makes excellent hay (provided it is handled and cured properly), and is a reliable, rich source of pasturage. Of equal importance, its capacity to produce nitrogen through its dense, nodule-laden root system makes it an ideal rotation and green manure crop. Often called "double cut" clover (mammoth red clover is called "single cut"), it can be counted on to produce two hay crops per growing season. It is commonly grown where rainfall is ample or where irrigation is available. Best growth is achieved on sweet silt loam or sweet clay loam soils, but it will tolerate more acid in the soils than will alfalfa. The major drawback to medium red clover is its short life. A biennial, it normally has a value for hay or pasturage for one year only. It is best sown in the early spring with a harvestable crop of small grain, or it is commonly intertilled with a previous crop of corn. When planted with companion crops like grain, the rate of sowing of the grain should be reduced by half to give the clover seedlings a competitive chance. The seeding rate for medium red clover is 10 lbs. p.a. unless mixed with grasses or other legumes, and it is important to inoculate the seed beforehand unless the acreage has successfully grown clover two to three years before. Hay yields commonly run 2 or 3 tons per acre, and the first cutting should be made when the blossom is in early bloom. Second cuttings should be delayed until the blossoms are in full bloom. If the crop is a healthy, pure stand, the weather is agreeable, and the plants have been visited by an adequate number of bees, the seed yield can amount to 4 to 5 bushels per acre.

4. Alfalfa. Indisputably the most important forage crop in the United States, alfalfa is a legume of high yield and considerable nutritional value. It is

Timothy

Brome Grass

Wheat

Alfalfa

Oats

Clover

adaptable, and (with local exceptions) will grow in all but the hottest, most humid areas of the country. An extensive taproot system that sometimes penetrates 30 feet into the ground makes it a good choice for hot, dry country having deep, well-drained soils. It is, however, intolerant of large amounts of alkaline salts (a problem in some western states), and it demands a neutral to slightly alkaline pH for sustaining yields. Despite their deep taproot systems, alfalfa plantings should be limited to short rotations in the driest sections of the Midwest (unless irrigation is available), and they should be discarded as an alternative in isolated sections of the North, where they have a history of winter kill. Alfalfa should be sown in late August or late fall in the Southwest, late spring or early summer in the Pacific Northwest, late summer or early fall in the South, and early spring in the North and Northeast. Unless alfalfa or clover has been grown on the land in the past two or three years, inoculation of the seed is necessary. It may be planted with a companion plant(s), and the rate is usually 8–14 lbs. p.a. in the East and 10–13 lbs. p.a. in dry areas of the West and Southwest (15–20 lbs. p.a. in irrigated sections). Hay yields vary considerably depending upon the number of cuttings (there are as many as eight cuttings of a stand that is irrigated in the West and Southwest), but 2 to 3 tons per acre is a working average on stands that see two cuttings per year. Assuming two cuttings, studies show that the best mowing time is when the plants are at the full-bloom stage. Successful seed production has historically been limited to the semiarid regions of the West and Southwest, and yields are 2 to 10 bushels per acre.

E. Equipment needed:

The following list is based on the least amount of mechanization needed to fertilize, till, cultivate, and harvest 20 or more acres of the grasses, clovers, and legumes cited above.

1. **Tractor.** This power source should have a potential of at least 30 horsepower, and have a power take-off (PTO).

2. **Spreaders.** Manure can be spread by hand, but larger acreages may require a manure spreader. Lime, rock phosphate, and other powdered fertilizers and amendments can usually be handled with a lime spreader.

3. **Seeder, drill, and roller.** With practice, these seeds can be spread by hand, but the small investment required for a spinner-type ("cyclone") seeder is warranted. If the practice requires it, a seed drill is unavoidable. Most grass, clover, and legume crops require rolling after seeding, and a simple cylinder-type roller or cultipacker should be used.

4. **Tillers.** A plow or rotating tiller is essential for breaking sod, but it may be replaced in some practices by a double-gang set of disk or bog harrows.

5. **Harrows.** Seedbed preparation requires a disk harrow, and occasional practices may involve a spring-tooth harrow. A smoothing harrow, though not essential, is a good tool for covering broadcast seed.

6. **Chopper.** This implement, fitted with a grass head, is necessary if the intended use of the crop is silage.

7. **Mower.** Harvesting hay in this quantity is best done with a sickle-bar or rotary-type mowing machine that is powered by the tractor PTO. If the crop (and the farmer's budget) warrants it, a haybine (a combination mowing machine, crusher, and windrower) can be bought to simplify the mechanical process. A haybine, or at least a crusher, may be necessary implements in areas where climate makes hay hard to cure.

8. **Rake, tedder.** Windrowing with a dump rake is satisfactory if the hay is to be put up loose, but dump rake windrows are often too heavy to be handled comfortably by a baler. With a baler, a side-delivery–type rake is the customary implement to use. Tedders can be employed in regions where hay is hard to dry, but with crops where leaf shatter is a problem the rake's gentler turning action should suffice to fluff the hay.

9. **Baler.** This implement is an essential one when the crop is to be sold or when storage space is limited.

10. **Combine.** Taking seed as a cash crop makes a small combine a necessity.

F. Field practices for timothy, brome, medium red clover, and alfalfa:

1. **Fertilizing.** All of these grasses need liberal amounts of calcium, phosphorus, and potassium. Manure, while a good general fertilizer, needs to be supplemented with phosphorus (amounts depend upon the results of soil tests). Manuring at the rate of 6 to 12 tons per acre is not uncommon. Sour soils

should be brought to a pH of 6.5 by the application of lime, and if the intent is to grow alfalfa (main or companion crop) and the soils are boron-deficient, borax should be added at the rate of 20–40 lbs. p.a. (**Note:** Borax can injure seed, and should be applied only as a top dressing on established fields, or worked into the soil at least a week before seeding.)

2. **Preparing the seedbed.** Following fertilization, the soil is worked into condition to receive the crop seed. The following is an outline of the general procedure for all four crops:

a. Plow or rotary-till fertilizer and/or soil amendments into the ground along with surface trash from the previous crop. Given the right soil, previous crop, and intended new crop (such as a green manure crop of clover following corn on light loam soils), the farmer may forego this step and proceed directly to disk harrowing. Lighter soils that can be easily wind-lifted or eroded should not be put to the plow until the last moment. In general, plowing need not be deep (4 to 6 inches is sufficient), unless one is plowing in a green manure crop or otherwise conditioning the topsoil.

b. Remove all stones that would interfere with subsequent practices.

c. Disk harrow the field. If sod is being broken, it may be necessary to break the clods with a spring-tooth harrow before applying the disks. Dead furrows from the plowing should be evened out, and this may entail overlapping or crosshatching coverage, but care should be taken to avoid refining soils (particularly heavy ones) too much. Powdery seedbeds tend to dry out too rapidly and are apt to crust over in warm weather following a rain.

3. **Seeding.** Careful choice of seed is essential, because some varieties are better suited to particular areas than others. Winter hardiness should be sought for seed to be sown in northern areas, and resistance to mildews and disease in southern areas. Certification for purity and germination should always be checked.

a. These crops are seeded by broadcasting or by drilling. Depth of planting depends upon the presence of moisture in the soil. Alfalfa plantings in arid regions of the Southwest, for example, are sometimes planted as deep as 1½ inches. The normal depth is ¼ inch to ¾ inch deep. Medium red clover, brome grass, and timothy grass are usually planted at about ¼-inch to ½-inch depths. If the seed is broadcast by hand or with a spinner-type broadcaster, it is usually covered by running a shallowly set smoothing harrow (spike-tooth) or a slightly cocked disk harrow over the surface. Light seeds, like brome grass or timothy grass, should be covered as soon after

Mowing/Combining a Field

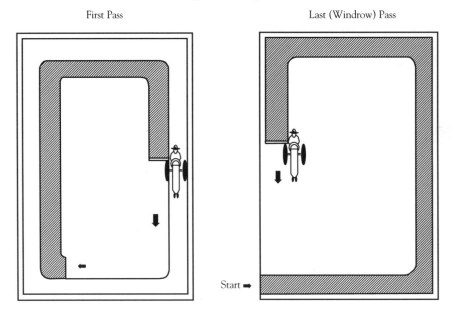

This method is advised for fields with uneven edges.

seeding as possible. Seed drills usually space the seed rows too far apart, and you must crosshatch the seeding pattern. Care should be taken to keep the seed mixtures (particularly those including brome grass) mixed in the hoppers of the seed drill.

b. After seeding, firm the seedbed with a roller or cultipacker.

4. Harvesting. These crops lend themselves to haying, chopping for silage, pasturage, or harvesting for seed. Silage crops are generally harvested at an earlier stage than hay. Timothy is taken slightly before the bloom stage, and alfalfa and medium red clover are taken anywhere from half- to full-bloom stage. Haying one of these crops is a three-step process involving mowing, curing, and gathering. The old adage, "Make hay while the sun shines," particularly applies to the small-scale farmer, as his operation generally is not large enough to warrant the use of driers or dehydration agents.

a. Mowing should begin when the farmer anticipates a string of four or five drying days. This is particularly important when harvesting clover or alfalfa. Curing of these two crops can be hastened by one or two days by running the crop through a crusher. If the mowing is the second (final) cut of alfalfa in winter-kill–prone areas of the North, the timing of the cutting is exceedingly important. Alfalfa needs to store food in its roots before the ground freezes, and a cutting made too late in the fall robs it of that food. As a rule of thumb, alfalfa should stand 12 to 14 inches above the ground as it goes into winter.

b. Curing hay properly in the field is a combination of good weather and good judgment. After the crop is mowed, haybined, or run through a crusher, it should be left in the swath until it is well wilted. Then it should be raked into a windrow. The raking should be done in the same direction that the field was mowed, and the number of swaths used to make a windrow depends upon the capacity of the baler and the thickness and height of the crop stand. Because of the potential for leaf shatter (especially with clover and alfalfa), subsequent rakings or teddings to fluff the hay should be kept to a minimum. Fluffings are necessary after a rain, and some leaf loss (10 to 15 percent is average) is unavoidable, but careless tedding can cost the farmer as much as half his crop yield.

c. Gathering the hay should be done only after it has thoroughly dried. The leaves and stems should still retain their natural green color, and they should retain their natural fragrance. It should be dry to the touch, and both the leaves and stems should be somewhat brittle. Whether loose or baled, hay should not be gathered until the morning dew has evaporated. In general, loose hay can be stored at a higher moisture content than baled hay. This advantage, plus the minimal mechanization required, appeals to many small-scale farmers. But baled hay takes up roughly half the space of loose hay, and hay can rarely be sold unless it is baled. Although there are baler attachments for automatic loading, the standard method is to leave the bales in the field behind the baler, and then gather these later. If these bales are left overnight, they should be turned to allow the bottom side of the bale to dry before picking them up.

d. Taking seed is a job for a combine. The seed heads should be taken when they are barely mature enough to thresh out when rubbed between the hands, or, as is the case with alfalfa, when three-fourths of the pods have turned brown or black. Letting the seed head become too dry will result in premature threshing. The proper time to take seed is early in the morning or late in the afternoon when the dew is on the head. Adjusting the cylinder speeds, choosing the correct sieves, and adjusting the air flow so that these small and lightweight seeds are not blown out with the chaff is a matter of carefully following the instructions that come with the combine. All four of these seeds should be dried and cleaned before being sold or put in storage.

II. ANALYSIS OF CORN AS A CASH CROP

Corn is an ancient grass from Central America that has, through selection, become the grain foundation of American agriculture.

A. General considerations:

Two types of corn are considered here: dent corn and sweet corn.

B. Purposes for growing this crop are:

1. Indirect cash crop. Dent corn, used as grain or silage feed for the farmer's own livestock, falls into this category.

2. Direct cash crop. Sweet corn is the primary direct cash crop for the small-scale farmer. Seed is not discussed here as a cash crop because of the specialization required to hybridize.

C. Rotation:

Corn is a heavy nitrogen user, and so it is customarily put into rotations following clovers or legumes. Bottomlands flooded annually can afford longer rotations of this crop.

D. Suitability (by type) of crops:

1. Dent corn. Hybridized varieties of dent corn are numerous, and the farmer should check with neighbors and local SCS and County Agricultural agents for the varieties best grown, for the purpose he intends, in his area. It is significant that nearly two-thirds of all the dent corn grown for grain in the United States is grown in the Corn Belt. Some regions, because of soil or climate, are just not suited to dent corn as a grain crop. While corn will produce a crop in any state in the country, slight variations in climate (cold nights at critical growth times, early freezing temperatures, etc.) can reduce yields by half. As a result, farmers in northern areas where early frost and delayed maturation are a problem take their dent corn as silage. This corn needs a frost-free growing time of at least 90 to 110 days, plenty of warm weather, and a well-distributed rainfall. July and August are critical months in most areas, and rain or irrigation is most needed then to offset losses from transpiration and evaporation. Good tilth and high organic content in the soils give the soil a high water-holding capacity, which will often sustain the corn over periods of no rain. Planting times for dent (and sweet) corn vary from mid-March in the South to late June in the Far North. Maturation times for dent corn varieties vary from 80 to 160 days, most taking 90 to 110 days. It is important to get the corn into the ground as early as possible in the Far North.

Planting is in rows spaced 36 to 48 inches apart, with plant spacing in the row varying from 2 to 12 inches, depending upon the use to which the farmer plans to put the corn. As a general rule, dent corn intended for silage is planted more thickly (4 to 8 inches) than that intended for grain (6 to 10 inches), but dry climates or poor soils may indicate wider spacings. Corn for grain is harvested later than that for silage, and it should never be stored in cribs

if it contains more than 30 percent moisture. Yields of grain corn are 40 to 64 bushels per acre and one can usually expect 15 to 25 tons per acre of silage.

2. Sweet corn. To the consumer, sweet corn is a tasty vegetable, but to a farmer it is an immature grass seed. Much of the information already given for dent corn applies to sweet corn. Soil and climate needs are identical, and, like dent corn, the sweet varieties are hybridized for maximum yield, early maturation, and uniformity. Unfortunately, not enough attention is being paid in hybridization to the gene reservoir of disease resistance. Although some pure strains are available, most farmers find that their customers prefer the hybridized ones. Sweet corn contains a larger percentage of sugar per kernel than dent corn, and it is available in yellow, white, or mixed varieties. Studies show that the yellow corn has more vitamins than white, and that white, generally speaking, has more sugar. Row-to-row spacing for sweet corn is the same as for dent corn, and in-row spacing varies from 8 to 12 inches, depending upon climate and the richness of the soil. Most farmers who market their corn directly to the consumer make it a practice to either space their planting dates so that the corn ripens at stages, or plant all of their crop at once using strains having progressively longer maturation times. From most sweet corns, one can expect yields of from 35 to 45 bushels per acre.

E. Equipment needed:

1. Tractor. A 30-horsepower tractor with PTO is the minimum needed for harvesting dent corn, and a minimum of 20 to 25 horsepower with PTO is needed for sweet corn.

2. Spreaders. For spreading fertilizers and/or amendments on acreages exceeding, say, 10 acres, a manure spreader and a lime spreader are essential.

3. Seeders. A corn planter or lister planter (depending upon local conditions) is indispensable. If intertilling clover between rows of corn, a spinner-type seeder is useful.

4. Tillers. A plow or rotating-type tiller is necessary in most corn enterprises, because corn usually follows a sod crop in rotation.

5. Harrows. Disk harrows and spring-tooth harrows are used to break up sod and to refine seedbeds.

6. Cultivators. Tractor-mounted, two-row

cultivators should be used if herbicides are not employed.

7. Chopper. Fitted with a corn head, this implement is used in making silage.

8. Corn husker. This machine can be replaced on smaller plantings with hand husking, but it becomes necessary with 10 or more acres.

F. Field practices for dent and sweet corn:

1. Fertilizing. It is seldom a question of whether to fertilize corn, only one of how much. If it is available, manure (6 to 10 tons per acre) is the most common fertilizer, otherwise liquid commercial fertilizer is added at the time of planting. With the former, it is necessary to add phosphorus. Corn often indicates its troubles with clear color signals: The lack of nitrogen produces a yellow discoloration of the older leaves that shows itself first at the tips and later along the midrib of the stalk. Purple discoloration of the stems and leaves of young plants indicates a phosphorus deficiency. A yellowing around the edges of leaves (as opposed to the midrib) indicates a lack of potassium. Rates of application of lime and other soil amendments depend upon a soil test, and if needed they should be added at the time of planting.

2. Preparing the seedbed. Plowing the seedbed is the usual first step for corn, but in some areas where soils permit, corn follows corn (or another open-cultivation crop), and one can simply disk under the fertilizer, amendments, and trash residues before planting.

    a. Plowing or tilling. The intent of plowing or tilling is to bury the sod and mix the fertilizers into the topsoil. Plowing or rotary tilling is usually done in the spring, but in areas where a short growing season dictates it, one can plow in the fall. Rotary tilling in the fall on light soils is not recommended, because it leaves the surface too smooth and encourages erosion in the spring. Fall plowing is particularly advantageous in level areas where the soil is heavy or where the land is wet in the spring. With fall plowing, the land should be left rough (undisked) over the winter. Obviously one should not fall-plow erodable, sloping land, or land where sandy soils are prone to wind lifting. Not so obvious is the fact that it does not pay to fall-plow first-year growths of biennial clovers. These clovers may well survive and become a problem in the corn

the following year. Spring plowing will effectively kill biennial clovers.

    b. Remove all stones. A second stone removal may also be required after the first harrowing.

    c. Harrowing. On most soils a thorough disk harrowing will complete the seedbed preparation. Crosshatching or lapping coverage depends upon the roughness of the surface. Stony soils or rough, hard-to-break clods may require the use of spring-tooth harrows. *Do not* reduce the seedbed to a floury condition.

3. Planting. Because corn is more prone to frost damage, it is usually planted later than small grains in all areas except the Far North. Northern farmers deem fall frosts more risky than late spring frosts. Careful row-to-row spacing cannot be overemphasized if the farmer intends to cultivate his corn. Using a two-row planter or lister with a row marker makes this spacing easier. Two-row surface corn planters are used in most sections of the country, but the two-row lister planter is still employed in arid reaches of the West and Southwest. Lister planters have small plows that precede the planter. These plows throw soil to both sides, and the following planter then spaces the seed in the bottom of the furrow. As the plant grows and is cultivated, the soil is pushed back into the furrow, leaving the roots at a deeper, moisture-favoring depth. As a rule of thumb, heavier planting rates produce more leaf, and lighter ones more grain. Planting depths are 1 to 4 inches, depending upon the soil and climate.

4. Cultivating. This practice is of particular interest to organic farmers, as it replaces the use of herbicides. Because most corn growers now use these chemicals, cultivators have become archaic curiosities. Tractor-mounted cultivators are just one of many types available, but their common purpose is to control weeds and aerate soil. Tipped with various-shaped working ends, they are usually ganged in groups of three or four to work two rows at each pass. Spring-tooth cultivators are favored with lighter soils, and the ridge-tooth type with heavier ones. Three cultivations per crop is a standard practice. The first is done when the corn is 2 to 3 inches tall, the second when it is 8 to 12 inches, and the third when it is 16 to 20 inches. Most farmers prefer all shovel tips for the first cultivation; replacing a shovel with a half-sweep nearest the corn on the second; and half- and full-sweeps for the third. The reasoning behind these changes of tips is to regulate the cultivating depth so that the corn roots are not damaged. The best cultivating depths are: 3½ inches, 1½ inches, and 1½ inches for the first, second, and third cultivations. Since the distances between the cultivator gangs are usually set to match the row-to-row spacing, it is obvious that errant planting can result in considerable crop loss during cultivation.

     5. Harvesting.

        a. Corn for grain is harvested when it has cured in the field so that the moisture has dropped to 15 to 20 percent. Corn should never be stored if its moisture content exceeds 30 percent. Grain that is sold on the wholesale market is graded on the basis of its moisture content — the less moisture, the better the grade. Moisture meters are the surest method of determining when to harvest. Once harvested, corn that is to be shelled and milled should be cribbed for storage. Eared corn requires 2½ cubic feet of storage space per bushel, and crib widths should be 7 feet in the Corn Belt, 5 feet in the North, and 9 feet in the Southwest. Cribs are best constructed with wire mesh above a concrete floor. A 3-foot rodent protection composed of ½-inch mesh hardware cloth around the base of the crib is an expensive necessity.

        b. Corn that is to be used for silage is chopped while in the early dent stage, and then ensiled in standing or trench silos. Plans for constructing the latter are available from local County Agricultural agents. Silage is a common choice of northern farmers having livestock (usually dairy cattle) to feed. The stalk, leaves, and grain are chopped. Harvesting corn for silage is a desirable practice in rainy or short-season regions, because the corn may be harvested in almost any kind of weather. After chopping it is hauled to the silo site and dumped or blown into place. Fermentation is the preserving factor with silage, and this process is dependent upon the action of anaerobic bacteria. These bacteria work only in the absence of oxygen, and it is therefore necessary to make the silo airtight. Plastic is used for this purpose with trench or bunk silos.

        c. Harvesting sweet corn is a non-mechanical operation for small-scale farmers. Hand-picking by the farmer (or by the consumer in "pick-your-own" operations) is indispensable for selectiveness. "Fresh-picked corn" is a meaningful slogan in the case of sweet corn, because the corn's sugar begins to change to starch the moment it is stripped from the stalk. Within eighteen hours most of the sugar in sweet corn is converted to tasteless starch.

III. ANALYSIS OF WHEAT AND OATS AS CASH CROPS

A. General considerations:

     1. Wheat is mostly grown for human consumption, but some by-products are used for livestock. Wheat yields vary from 20 to 40 bushels per acre, depending upon the variety, climate, and soil conditions.

     2. Most oats are grown as feed for livestock. Yields for oats are 40 to 80 bushels per acre.

B. Purposes for growing these crops are:

     1. Indirect cash crop. Oats can be grown for grain (mostly livestock feed), or it can be cut as hay. Only 4 percent of oats grown in the United States are for human consumption.

     2. Direct cash crop. Wheat and oats can be sold directly to the consumer (an option that usually requires storage facilities), to a wholesaler, or to consumers or wholesalers as seed. Straw from either crop can be sold as bedding or mulch.

     3. Soil investment. Oats are often used as a nurse crop, and they are taken as sheet compost or as hay before they have reached their full term.

C. Rotation:

     Oats or wheat usually follow an open crop like corn or soybeans, but the introduction of a well-grown green manure crop of clover or alfalfa into the rotation before them can satisfy the nitrogen needs of both wheat and oats.

D. Suitability:

     1. Wheat. This crop is adaptable to all but the hottest and most humid regions of the country. It is tolerant of sandy loams and even clay soil, and will grow (with varying degrees of success) in areas having rainfalls ranging from 20 to 70 inches. Ideal soils for wheat are sweet (6.5 pH), fertile, and of silt loam or clay loam consistency. Five major varieties of wheat are grown in the United States. They are: hard red spring (North and Northeast), hard red winter (Midwest and West), soft red winter (East), durum (Northern Plains states), and white wheat (Far West).

Farmers should check local sources (SCS or County Agricultural agents) before finally selecting a variety.

a. Hard red spring wheat. Grown where winters are too severe to plant overwintering varieties, this wheat produces a superior bread-making flour.

b. Hard red winter wheat. Over half the wheat grown in the United States is of this variety. Sown in the fall, this wheat makes a flour almost equal to the hard red spring variety.

c. Soft red winter wheat. This grain is starchier and has less gluten than either hard red winter or hard red spring wheats. In blends with the above, this grain makes good bread-making flour.

d. Durum wheat. Used in the manufacture of flour for pasta, this grain has a limited market. Its weak gluten content makes it unsuitable for bread flour. Red durum wheat is grown in limited amounts for poultry feed.

e. White wheat. Both spring and winter varieties of this wheat are grown in the Far West. Small quantities are also grown in Michigan and New York. Flour made from this grain is suitable only for pastries.

2. Oats. Over 90 percent of the oats grown in the United States is for livestock feed, and unless the small-scale farmer has expensive hulling and rolling machinery or a contract with a middleman, he should stick to growing livestock feed. Among the many types of oats available, two dominate in common usage: common oats and red oats. The latter are adapted to warmer climates and the former to

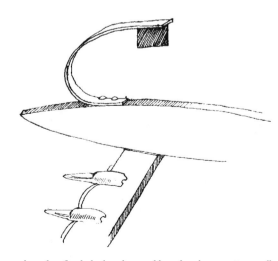

A metal marker flag bolted to the swathboard makes mowing a tall grass crop easier.

cooler. Of the two, common oats are much preferred because they return greater yields in grain and straw, and they contain a lower percentage of hulls. Oats are easy to plant because they require little advance work. Their seedbed need not be carefully manicured, and adding fertilizers before planting only causes the straw to grow rank and lodge (fall over) before the grain is ripe.

E. Equipment needed:

1. Tractor. The combine needed to harvest this crop makes a tractor of at least 30 horsepower a necessity, and the tractor should have PTO.

2. Spreaders. As with other grass crops, a manure spreader and a lime spreader are needed for larger acreages.

3. Seeders. Oats or wheat can be drilled with a seed drill, or they can be broadcast with a spinner-type seeder. In an emergency oats can be distributed with a lime spreader. Though not mandatory, it is a good practice with either oat or wheat seedings to roll or cultipack the seedbed; therefore one of these implements would be necessary.

4. Tillers. Depending upon what precedes these grains in rotation, a plow or rotating tiller may be essential.

5. Harrows. A disk harrow is necessary with all grass crops, and occasional circumstances may require a spring-tooth harrow. Where a seed drill is not available, some farmers drag a spring-tooth harrow in a pattern to make small furrows and ridges. They then broadcast the seed, the seed tends to fall into the furrows, and subsequent rolling or cultipacking covers the seed into the harrowed rows.

6. Combine. This implement is essential for taking grain from these crops. Combines that have a 5- to 7-foot cutting swath will handle most needs of a small-scale operation.

7. Rake. If straw is to be marketed, it is necessary to windrow the swaths for curing and baling.

8. Baler. As with haying operations, straw can be stored loose, but if storage space is a limiting factor the straw should be baled.

F. Field practices for wheat and oats:

1. Fertilizing. It is usually best to add nitrogenous fertilizers like manure on the crop that precedes the wheat or oats. The reasoning is that most of the nitrogen will have been used in the old crop, and the residues are sufficient to grow these cereal grasses. But in the case of fall plantings, an extra dose of nitrogen in the spring (like well-rotted manure) is welcome. Lime, should it be necessary, should be added at this time, and long-release fertilizers like rock phosphate can also be worked into the soil in the spring.

2. **Preparing the seedbed.** When oats and wheat follow corn, soybeans, or other open-cultivation crops, it is common only to disk the field before seeding. The process of plowing, stoning, and disk harrowing is necessary if sod crops precede wheat and oats in the rotation. Fall plantings require plowing as soon after harvest of the preceding crop as is possible, and in some areas it is a good practice to fall-plow for a spring planting. In spring planting, follow the procedures outlined for clover and alfalfa. As a final step before drilling, it is a good practice to roll or cultipack the seedbed, but this step should be omitted if the soil is too wet.

3. **Seeding.** Most wheat is sown at the rate of 5 to 8 pecks p.a., and oats at the rate of 8 to 10 pecks p.a. However, arid sections of the West and Southwest will not sustain this rate, and it should be halved. Oats that are planted as a nurse crop should be planted at half the above rate. A seed drill is favored for both wheat and oats, but both can be broadcast. If broadcast, one should add 15 to 20 percent more seed. Planting depths for wheat or oats vary from 1 to 2 inches in the more humid areas to 2 to 3 inches in the drier sections of the country. A uniform depth is difficult to achieve with broadcast seed, as the covering is usually done with a smoothing harrow or a roller, or both. It is a good practice with either of these seeds to get them into the ground as early as possible, in the fall or the spring, and it is reassuring to know that either seed will germinate at temperatures below 40°F. (5°C.).

4. **Harvesting.** Nurse crops of oats should be taken for hay or chopped feed as soon as the main crop is firmly established. This is done with a mowing machine or a chopper with a grass head. Some care should be taken during this operation to leave the main crop as undisturbed as possible. Raking and baling will not usually set the crop back significantly if the young plants are thrifty. Mature oats and wheat are harvested when they reach the hard dough stage. This is characterized by the heads turning yellow and the naked grain being hard but dentable when pressed with the thumbnail. A good test for threshability is to rub the heads to see how readily the grain releases. No stand of wheat or oats will ripen exactly at the same time, and the farmer should exercise some judgment in deciding when he will get the most grain with the smallest losses to shattering. If he has a moisture meter, he can go about this task more scientifically. These grains should not be stored if the moisture content exceeds 14 percent. Field losses can be reduced by careful settings of the combine, and processing and storage losses can be lessened by subjecting the grain to air-drying. If a heat-augered bin is not available, suspect grain should be spread out or put in a fine-meshed wire-bottomed bin that has provisions for air circulation below. Frequent turning of the grain with a shovel will insure even drying. If the grain appears dry enough, it may be stored in porous bags (out of the reach of rodents), but the farmer should make daily checks for signs of mold or heating.

CHAPTER 6

Other Cash & Specialty Crops

Choosing among the multitude of other cash crops, the beginning small-scale farmer has to consider his crop rotations, his markets, and his pocketbook. The scale of his operation will eliminate many possibilities, and climate and soil conditions will narrow his options even further.

Potatoes, dried beans, raspberries, and strawberries are discussed here because they offer a range of crops from which almost any farmer can choose. More important, these vegetables and berries are suited to small-scale ventures by virtue of their labor intensiveness, and their sure (if not excellent) market potential.

Potatoes and beans are accepted parts of the American diet, and because of the increasing prices for meat, protein-rich beans promise to increase in consumer popularity. Though the farmer should have a ready-to-buy market for his potatoes or beans when he harvests them, circumstances may require that he store them for a short period of time (or even over the winter). It is comforting, and sometimes profitable, for the farmer to know that he does not have to move his crop to market immediately.

But berries take the blue ribbon for consumer demand. Nutritionally bankrupt, compared to vegetables like beans and potatoes, raspberries and strawberries are nevertheless the nearest to a "sure thing" that the small farmer can market, provided his prices are competitive and his product known.

Raspberries and strawberries are particularly attractive as a cash crop to those farmers having limited acreages. They have a high profit-per-acre capacity, and are eminently suited to making profitable use of those awkward-to-work parcels that seem to occur on most farms. Because these berries are both labor-intensive and perishable, they are not especially favored by large commercial growers, but they are well suited to the small-scale farmer.

I. ANALYSIS OF POTATOES AS A CASH CROP

A. General considerations:

1. **Size of enterprise.** Anywhere from 1 to 10 acres is a good size for the small venture, but the acreage one selects is critical insofar as equipment is concerned. Hand planting and hoeing are still feasible practices for raising 1 or 2 acres of potatoes, but plantings of 3 or more acres are more sensibly handled with tractor-drawn equipment. Older, outmoded one-row equipment is usually available, and some items (like hillers) can be tinkered up to meet the small farmer's needs.

2. **Soil and climate.** Potatoes of one variety or another can be grown almost anywhere, but they do best on sandy or gravelly loams. Planting them in heavy soils that are poorly drained is asking for trouble, and alkaline or neutral soils will invite scab. If the soil is already of good tilth, and has a good supply of organic material, the usual step of planting a legume green manure crop before putting in potatoes may be omitted. Potatoes require a large amount of water (the tubers themselves are 75 to 80 percent water) early and late in their growing cycle, but they can be grown on dry lands if certain circumstances exist. There should, for example, be a depth of 3 feet or more of moist soil at planting time, and the crop should be preceded by a fallow year in the rotation schedule. Although some potatoes are grown in the Southeast, they do much better in the cooler climes of the Northeast. USDA figures support this generalization by showing that potato production decreases as the temperature and length of growing day increases.

3. **Markets.** Over 80 percent of the nation's crop of potatoes is consumed by humans, and this is the market at which the small-scale farmer should aim. While growing seed potatoes can be profitable, particularly with the rarer local varieties, the storage factor involves expense and a risk of spoilage that the small grower should leave to others. Planting to fulfill advance orders, with growing contracts, is the surest marketing procedure, but many small-scale farmers have found an on-farm, dig-your-own operation successful.

B. Rotation:

Soil-borne diseases (scab, fusarium rot) dictate that potatoes should never be planted on the same land more than once in three years. The method of rotation varies from region to region, but as a general rule, potatoes do well on a new breaking of sod. Most success is achieved by having them follow legumes like alfalfa, sweet clover, medium red clover, or peas, but some regional rotations find potatoes preceded by corn or beets. After consulting with the County Agricultural Agent, to determine the rotations common to the local area, the small-scale farmer should keep from committing himself until he has assessed the last-minute weather and economic conditions that will affect his crop and market.

Potato Plant

C. Varieties:

After their wild beginnings in the high Andes, potatoes were carried to Europe with the conquistadores, and thence to this country with the colonists. Their wanderings have produced thousands of adaptations to meet the climatic circumstances into which the potatoes were thrust, and the tuber we plant today as a cash crop has no resemblance whatever to the ancestral, knobby, misshapen tubers that are still offered for sale in the markets of La Paz. The four most popular early varieties we now plant are Irish Cobbler (round, white), early Ohio (oblong, pink, with corky dots), Norland (red-skinned), and Triumph (round, light red skin). Two of the most commonly planted midseason varieties are the Kennebec (round, white) and the Superior (round, white). Of the late-maturing varieties, growers favor Katahdin (round, white), Green Mountain (oblong, white), Rural (round, white, mostly grown in regions having hot, dry climates), and Russet Burbank (long, spindle-shaped, mostly grown in the West).

D. Fertilizers and amendments:

As was pointed out earlier, potatoes are tolerant of acid soils. In fact, acid soils (4.8 to 5.4 pH) are sought after in the more humid areas of the Northeast to help control the fungus that causes scab. Therefore, lime should not, as a rule, be used on a bed that is being prepared for potatoes, though it may be necessary to add it in order to grow any leguminous crop that might precede the potatoes in rotation. A soil test should be made the year before planting to determine the NPK (nitrogen, phosphorus, potassium), pH, and trace mineral needs (potato yields can be increased substantially in some areas by the addition of small amounts of magnesium). In most instances, an application of cow manure (horse manure promotes scab) of 6 to 10 tons per acre on the crop preceding the potatoes is considered sufficient for the following year's crop as well. Heavy applications of cow manure the year of seeding, unless the manure is well-rotted, should be avoided, as this also promotes scabby potatoes. At the time of the manuring, it is advisable to also add phosphate in the form of superphosphate or rock powder. The usual amounts of this supplement vary, but 50 to 75 pounds of 20 percent superphosphate or 100 to 150 pounds of rock phosphate per ton of manure is usually sufficient. This amount of rock phosphate will not yield the same percentage of *available* phosphate, but will release its goodness over a protracted period of time.

In summary, organic growers usually rely on the green manure crop and residual nutrients from the previous year's fertilization for the needs of this year's crop of potatoes. Growers of the chemical persuasion often use the same practices, but also add a "complete" chemical fertilizer at the time of planting (NPK amounts vary depending upon soil needs). This fertilizer is placed in two bands — 2 inches to the side and below each seed piece.

E. Equipment needed:

After seedbed preparation, 1- or 2-acre plantings can be handled with hand tools ("jabber," hoe [or walk-behind tiller], shovel, potato fork). For seedbed preparation, seeding, cultivation, hilling, and harvesting of 3 to 10 acres you will need:

1. **Tractor.** Twenty to 30 horsepower is sufficient for most practices outlined here. A PTO is helpful, but not usually necessary. Walk-behind tillers can be used on smaller acreages.

2. **Spreaders.** Depending upon the size of the enterprise, a manure spreader and a lime spreader may be required.

3. **Seeder.** If the area is less than 2 acres, the grower can manage without a mechanical seeder, but if he can acquire an old-fashioned, one-row planter (having opening disks in front of the delivery tube, and closing disks after) the process is much simplified, and the coverage generally more even. These units are tractor-drawn, and typically require an extra person to ride herd on the seeder.

4. **Tillers.** A 14-inch plow or a rotary tiller is essential to prepare the seedbed. The plow may also be used to open seeding furrows.

5. **Harrows.** Disk harrows are necessary if the land is plowed.

6. **Cultivator.** A one- or two-row cultivator is needed for weed control of larger plantings.

7. **Hiller.** A one-row disk hiller is needed on larger plantings.

8. **Potato digger.** A one-row, elevator-type digger is a welcome addition when harvesting larger acreages.

9. **Sprayer.** If insecticides are used, a shoulder or backpack-type sprayer is the minimum amount of equipment needed.

F. Field practices for potatoes:

1. Seedbed preparation.

a. **Plowing.** In general, heavier soils benefit from fall plowing, lighter soils do not, but in either case cover crops or manure should be turned under early enough for the organic material to rot thoroughly. A deep plowing (8 to 12 inches) and thorough disking is essential prior to planting, but if the topsoil is shallow, it does not pay to bring up large amounts of subsoil. Having quantities of trash to be turned under may require a 16-inch bottomed plow. The deep plowing is necessary both for soil friability, and to bury the recommended green manure deep enough to discourage volunteers.

b. **Stoning and disking.** All stones larger than a baseball should be removed. A single coverage disking will usually discover additional stones, and, after they are hauled off, the field should be lap-covered with the disk harrows at right angles to the plowing direction.

c. **Leveling.** If irrigation is planned, the land must be graded so that the water flows uniformly over the area. The best grade for potatoes is 5 feet of drop per 100 feet of row, and provision must be made to drain the lower ends of the crop rows. Potatoes will not tolerate flooding.

2. **Seeding.** The most important single factor in a successful seeding practice is to buy seed that is healthy, firm, uniform, and *certified*. Employing all the best farm practices in the world will not offset putting diseased seed into the ground. Commercial growers further reduce their disease potential by bathing their seed in a poisonous dip of mercuric acid or formaldehyde solution. Organic growers who are careful in their rotational plan count on keeping soil-borne disease organisms to a minimum, and thereby bypass this disinfecting process.

Until ready to plant, seed tubers should be stored in a cool (36° to 49°F. — 2° to 10°C.) dark room. Before cutting, they should be exposed to light and warmer temperatures (60° to 70°F. — 16° to 21°C.) for "green sprouting." The seed tubers should be spread out and turned every four or five days. Sprouts should not be allowed to grow more than 1 inch in length, and they should be stoutly formed. Planting sprouted seed can hasten maturity by as much as two weeks.

If the seed is cut into blocky chunks before planting, 100 pounds of seed can be expected to cover 400 to 500 feet of row. The cutting should be done with a sharp, thin knife; the pieces should weigh between 1½ and 2 ounces per chunk; and each chunk should have at least one sprouted eye. Timing the sprouting and cutting operation to coincide with an ideal planting time is a tricky proposition, but one can gain time by putting the seed back into the lower-temperature, no-light room.

Whether planted by hand or by machine, seed potatoes should be spaced 8 to 10 inches apart in rows that are 32 to 36 inches apart (Russet Burbanks need 12- to 14-inch in-row spacing). The seed should be set 4 to 5 inches deep in moist soil except when irrigation is anticipated. In the latter instance, 2 to 3 inches of planting depth is sufficient.

Planting by hand can be made easier by using the plow to cut a shallow furrow. The seed is then spaced in the furrow, and the dirt put back with the plow or with a hoe.

3. **Weed control and hilling.** Some large growers use herbicides for weed control, but this practice is *not* recommended here. Two or three cultivations and hilling passes, beginning when the plants are about 5 or 6 inches high, are the surest and safest method of weed control. As the potato plants grow, they should have soil thrown up on their exposed stems. This practice prevents sun scalding of exposed tubers, and controls weeds next to the plants. Between-row cultivation should employ shovels or sweeps, but care must be taken to avoid pruning the plants' roots. This is done by keeping disks or half-sweeps that are nearest the row well back from the plants, and never cultivating deeper than is necessary to kill the weeds. In irrigated fields, the first cultivation is usually followed by ridging with a winged (or disk) hoe. This practice leaves channels between the rows in which the water may flow.

4. **Spraying.** Insects and fungi that bother potatoes are a perennial problem for organic farmers. Relying on low-toxicity dusts like rotenone, ryania, and Tri-excel, they are forced to make many more applications than the grower using an all-purpose chemical insecticide/fungicide. Chemical fungicides control early and late blights, and if he will not use them, the organic grower must ruthlessly rogue out (remove diseased plants from) his fields, and hope that he has controlled the spread early enough. No sprays or dusts should be applied until the plants are at least 5 or 6 inches high.

5. **Harvesting.** When the potato crop is harvested depends upon the climate and the farmer's market. Yields will increase with maturity, but timely higher market prices may induce the farmer to take his crop immediately after it gets to marketable size. Maturity is reached when the vines are dried-up and the tubers have hardened. Yields can vary from 200 to 300 bushels per acre, but if a small-scale farmer has really applied himself, he should not be happy unless he has gotten 400 to 500 bushels per acre.

The old-fashioned way to harvest potatoes is to use a shovel and a potato fork. This method can still be used on 1- or 2-acre parcels, but a single-row, elevator-type potato digger simplifies matters considerably. One other method is to set the plow deep enough to bring the potatoes to the surface, but not so shallow as to cut or bruise them. This practice is usually followed up by a careful raking with the potato fork. Once exposed, the potatoes should be allowed to dry, and then they are gathered in crates or burlap bags.

II. ANALYSIS OF DRIED BEANS AS A CASH CROP

A. General considerations:

1. Size of enterprise. Like potatoes, a workable size for a small-scale farm venture in dried beans is anything from 1 to 10 acres, but the similarity stops there. Growing dried beans is more akin to corn, and, in fact, employs much of the same equipment. With 1-acre enterprises, the farmer can get by without the most expensive piece of equipment, the combine. One should expect to harvest 800 to 1,200 pounds of dried beans per acre.

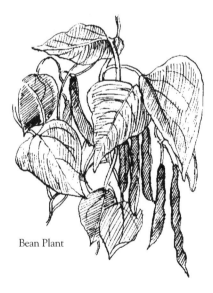

Bean Plant

2. Soil and climate. Most dried bean cultivars (man-perpetuated varieties) prefer friable, neutral sand or silt loam soils. Roughly speaking, they are grown on the same soils as corn. Though they seem to do better in the central and northern states, they can be grown in any region of the country. Irrigation, for example, makes them a desirable cash crop for the West and Southwest (except those areas having a large buildup of alkali salts). Areas having very short growing seasons (less than three months) may not lend themselves to this crop, for beans are intolerant of frost.

3. Markets. Because dried beans have only 12 to 15 percent moisture and are eminently suited to storage for periods of two or three years, they offer the grower an enviable range of options. He can sell them "behind the combine" or whenever the market offers the highest price. They are an excellent crop to offer to food conspiracies, or to other food-contracting organizations like co-ops. Like the general consuming public, these organizations are becoming sensitive to food that makes ecological sense, and beans are exemplary in their efficiency as a direct protein food source. Lastly, no bean grower should overlook the possibility of marketing his crop as seed.

B. Rotation:

As a legume, the bean plant is a welcome addition to most rotations. Beans typically precede corn in three- or four-year rotation schemes, but they should never be grown more than once or, at the most, twice in succession.

C. Cultivars:

There are nearly as many cultivars of beans as there are farmers growing them. Any kind of bean can be dried, and it would seem that most of them are. Many popular local strains of beans take on an heirloom character, and the small-scale farmer who wants to take advantage of the local market should seek these beans out for seed. Since beans are self-pollinating, these rarer strains are relatively easy to keep "pure."

Aside from the heirloom cultivars, dried beans that lend themselves to cash cropping can be comfortably grouped into two cooking categories: the stewing beans and the baking beans. Stewing beans include such well-known variegated beans as pinto, French horticultural, shelleasy, and red kidney (the kidney is also used as a baking bean). Examples of the more popular baking beans are: navy, flat Great Northern, Jacob's cattle, soldier, small pea, sulphur, and Michelite.

D. Fertilizers and amendments:

For the most part, beans provide their own nitrogen, but experienced bean growers plan their rotation in such a way that they can plow down a legume like clover in preparing the bean seedbed. Some lime may be required, but beans are tolerant of soils ranging from pH 6.0 to 6.8. Beans use up more phosphorus and potassium from the soil than does corn, and so fertilizers should include concentrations of these elements. Quantities vary, but typical commercial NPK mixtures are: 0–20–10, 0–20–20, and 0–12–12 at 200 to 300 pounds per acre. Rock dust fertilizers (rock phosphate, greensand, or granite dust) should be added in bulk quantities that are roughly double that recommended for commercial fertilizers. Also necessary to healthy bean growth is zinc. Adding 8 to 10 pounds of zinc sulphate per acre is a good practice where requisite amounts are suspect. This amendment should be done through the fertilizing attachment on the seeder rather than by broadcasting.

Organic fertilizers are too slow-acting to serve as banded starter, so most organic growers forego the commercial practice of adding a "complete" chemical fertilizer (10–20–10 or 10–20–20 at 200 to 300 pounds per acre) at the time of seeding. These fertilizers are usually applied in two bands placed 2 inches to the side, and below the bean seed. A corn planter does this job nicely.

E. Equipment needed:

The same equipment used for corn is required for beans, plus the following:

1. **Mower or bean cutter.** For drying, bean plants must be mowed or cut. A standard mowing machine will do the job if a bean cutter cannot be found.

2. **Combine or stationary thresher.** One-acre patches can be hand-threshed with a flail, but larger ones require a combine or a stationary "barn" thresher. The combine can be of the smaller variety. One that cuts a 5-foot swath will handle up to 10 or 15 acres. For ease of field handling, a combine should have a pickup-type head.

3. **Seed cleaner.** Also known as a "fanning mill," the seed cleaner further cleans and dries the beans after harvesting.

4. **Drying bins.** Air-drying can be done with simply constructed, ventilated wooden bins, or one can do the job with grain bins having heat augers.

5. **Bean sorter.** Hand-sorting beans on the kitchen table after supper is a traditional excuse for gossiping, but talk will run thin before 3 acres of beans have been sorted. Any amount over 600 pounds warrants the purchase or construction of a continuous-belt bean-sorting table.

F. Field practices for beans:

1. **Seedbed preparation.** The land should be fitted for beans in the same way as for corn.

2. **Seeding.** Using seed that tests less than 70 percent germination is a poor practice — most should germinate at 85 to 90 percent. The one- or two-row corn planter does a good job of seeding beans *if* the proper bean plates are available. The fertilizer cans on the corn seeder are satisfactory for banding fertilizer.

Seed should be inoculated with a nitrogen-fixing bacteria before planting. Inoculation is done just before seeding, by stirring the inoculant among seed that has been coated first with sticky sugar water. A cement mixer does an adequate job with larger amounts of seed.

Large commercial, nonorganic growers subject their seed to a dust bath of fungicide and insecticide before inoculation. Organic farmers substitute for this process by planting an additional 10 percent of seeds. The idea is to anticipate the losses to diseases and insects while the plants are sprouting and thrusting out of the soil.

Seeding rates are usually figured at 1 bushel (60 pounds) per acre, but larger beans will take more, smaller beans less. Row-to-row spacing is usually 32 to 36 inches, and in-row spacing is 3 to 4 inches. The seed should be covered by about 1 to 1½ inches of soil.

3. **Weed control.** Although most large growers of dried beans use herbicides, small-scale farmers will find cultivation more to their purposes. Three or four cultivations, as with corn, are all that is required, and they can be started quite early. The first is done as soon as the plants are 3 to 4 inches high, the second when the first weeds begin to appear, and the third before the runners are formed. To prevent the spread of plant diseases, cultivate when the dew is off the bean leaves. No cultivation should be done after the bean pods are formed.

4. **Spraying.** Sprays are preferred to dusts, and they include insecticides, fungicides, and herbicides. Time and rate of application depend upon the kind of problem the plants have, and the severity of the infestation or outbreak.

Organic farmers who refuse to use these poisons will rogue (cull out diseased or insect-ridden plants from) their bean patch on a weekly basis, but, unless these problems are discovered early, it is usually a futile effort.

5. Other pests. No description of the bean-growing enterprise would be complete without mention of the losses that can be suffered from the depredations of rabbits, other rodents, and deer. Extermination programs can keep the rodent damage to a minimum, but deer either have to be fenced out (prohibitively expensive), or the grower should consult local game officials as to any compensatory damage payments they may make for the deer's encroachments. Better yet, if deer are numerous enough to constitute a threat, the farmer should give serious consideration to planting a different cash crop.

6. Harvesting. The mark of maturity in dried beans is hardness. Too early harvesting will guarantee mold, and too late will cost the farmer a percentage of his yield when his beans are prematurely threshed. The thumbnail test is the usual field practice (hard-to-dent seed is mature), but testing with a moisture meter is surer. With some help, the farmer can pull his beans by hand, but larger acreages are more easily handled with a standard mowing machine or with an old-fashioned bean puller. Once the plants are loose, they are windrowed (with their roots turned up) for drying. Three or four consecutive dry days are required for most mature crops. Mowing or cutting should be done in the early morning while the dew is still on the plants. This practice will cut down the losses to premature threshing.

Threshing is the next step. If a stationary thresher is used, the crop has to be transported to the machine. The plants are then fed into the maw of two-toothed cylinders that revolve at different speeds. These teeth shatter the brittle pods and drop the beans and chaff onto air-blasted, oscillating screens. Combining involves the same process, except that the cylinders are wreathed with rubber-tipped beater bars, and the machine is mobile. Some combines are mounted with a pickup device that allows the operator to drive the machine along the windrow while an assistant rides on the combine to bag the beans. But if the combine has no pickup, the bean plants must be forked onto the combine's apron by hand. Threshed plants are returned to the field as trash to be incorporated at the next plowing.

7. Processing. Even the most efficient combines leave trash (stems, stones, soil, and split beans) in the beans, and they must be run through a seed cleaner. These cleaners do nothing more than repeat the air-blasting and screening of the threshing process. One or two passes through this machine may be required.

Because not all of the beans will be dry at this stage, they must be subjected to artificial drying. Air-drying is done in wooden storage bins having wire-meshed bottoms, under which fan-blown air is circulated. More affluent growers acquire large metal grain bins (typically 5- to 10-ton capacity), and dry the beans in them with electrically powered heat augers.

The last stage of processing is to hand-sort the beans — that is, to separate out the last stubborn pieces of soil or stones that evaded the cleaning process. "Hand-picked" beans are identified in the bean marketing world as the finest grade of beans (guaranteeing less than 1½ percent defects and foreign matter), and they are sold at the highest market prices. This laborious sorting can be made much easier by constructing a continuous-belt sorter from an old, treadle-type sewing machine. Old commercial models of this sorter can still be found.

III. ANALYSIS OF RASPBERRIES AS A CASH CROP

A. General considerations:

Specialized equipment is not a major consideration when deciding whether to grow raspberries. This bramble lends itself to labor-intensive, oversized gardening ventures. Some farmers have found ranging poultry in the raspberry plantation (for weed control and soil enrichment) a compatible mix of farm ventures.

1. Size. Two-person raspberry plantations, where no outside picking labor is employed, should never exceed an acre in size. Pick-your-own plantations can be larger, but beginners should not, at the outset, plant more than 2 or 3 acres — particularly if they do not have an established consumer clientele, or are not situated in a central marketing location. The following facts and figures should be considered by the farmer when planning the size of his enterprise:

a. One person can pick about an eighth of an acre of raspberries in one day.

b. In season, raspberries need picking every three or four days, and every two when the weather is hot and humid.

c. Yields average 1,300 to 1,500 quarts per acre, but this figure will double in top-notch years. Yields of Cuthberts planted in the state of Washington and in various eastern locations have run as high as 8,000 quarts per acre.

d. Raspberries are perishable.

e. Raspberries must be irrigated regularly if you expect high yields.

f. The productive life of a raspberry plantation is seven to ten years, but it does not reach its top bearing potential for three years.

Raspberry

2. Soil and climate. In siting his raspberry plantation, the farmer should give some attention to his soil type, but a handy supply of irrigation water and a place where there is good air and water drainage is more important. Raspberries of most types thrive on sandy loams having large amounts of organic material, yet equally good yields can be gotten on clay or sand soils if they are carefully managed. Some varieties, like the June, actually prefer clayey soil.

Raspberries prefer the cooler climes of the North and Northeast for their summer growth, but winterkill is a problem in the more severe wintering areas. The Pacific Coast states are historically successful places to grow these plants, and they can be grown under irrigation with moderate success in the Midwest and mountain states. The hot, humid summers of the South are not favorable to raspberry

cash crops, but they are possible if the plantation is carefully planned and sited (northern or northeastern slope). All raspberries need full exposure to the sunlight.

3. Markets. The best marketing situation that the small-scale farmer can create is one in which the consumer buys directly from him at this farm. On-farm marketing of raspberries eliminates spoilage or injury of the fragile berries due to transport, time lost in transporting the berries to market, and profit losses to middlemen. The perishability of his berries may well force him occasionally to seek outside markets (particularly when he is becoming established as a grower), but as a rule, the demand for these fancy dessert berries outdistances the supply. Raspberries are normally marketed by the pint. They are among the most time-consuming of berries to pick, and warrant a higher price.

The alternative to this is to establish a "pick-your-own" plantation. With this operation a parking area is necessary, and the farmer should supply uniform containers (at cost) so that the customer is sure of a full measure. Because pick-your-own operations seldom see the customers pick the canes "clean," it is necessary for the farmer to figure in some picking time to follow-up after his customers. These farmer-picked pints are usually kept in a refrigerator to fill out special orders, or for customers who, in the heat of a summer day, are ready to sacrifice the pick-your-own discount for convenience.

B. Rotation:

Because the raspberry plantation occupies space for seven to ten years, rotation is not usually considered a factor. This is a mistake. In planning the plantation it is important to orchestrate what crop precedes raspberries. Cultivated crops such as beans or corn are preferable to sodded crops. The key consideration in figuring preceding crops is weed control. Quack grass (also known as witch or Johnson grass) is the worst enemy of this berry crop, and clean cultivation or fallowing the area the year before are logical, weed-killing rotations. Other ways to kill quack grass are: planting and plowing several successive green manure crops of a quick-growing plant like buckwheat or oats; covering the acreage with black plastic for a growing season (expensive); eradicating by hand-weeding; or applying a herbicide. Hand-weeding guarantees the best degree of success if it is followed by a year of fallow. Because they are susceptible to the same wilt diseases, raspberries should never follow potatoes, tomatoes, eggplants, or peppers in rotation.

C. Varieties:

There are four different kinds of raspberries: red, black, purple, and yellow. Red raspberries are by far the most common, followed by black, purple, and yellow. The yellow are often included in a plantation for their novelty and for the attractive color contrast they lend to a picked pint. It is important for the grower to choose a variety within these color types that best suits his climate and soil. The following is a list of varieties commonly grown in a general geographic region. It is not meant to be complete, but should give the grower some guidelines when discussing the possibilities with another local grower, or the local agriculture agent.

North and Northeast

March (red)	Gatineau (red)
Latham (red)	Success (purple)
Earlired (red)	Allen (black)
June (red)	Bristol (black)
Indian Summer	Cumberland (black)
(ever-bearing red)	Amber (yellow)

Midwest

Latham (red)	Cardinal (purple)
King (red)	Amethyst (purple)
Cuthbert (red)	Forever Amber
Kansas (black)	(ever-bearing yellow)
Black Pearl (black)	

Far West

Surprise (red)	Marlboro (red)
Cuthbert (red)	Farmer (black)
King (red)	Cumberland (black)

D. Fertilizers and amendments:

On soils that are free of quack grass, one should apply heavy doses of cow manure, mulch straw, and whatever seed-free organic material there is available, to be plowed down the year before planting the raspberries. Eight to 12 tons of manure per acre is not too much, and 6 to 8 tons of straw per acre, though expensive to apply, is a welcome addition to

raspberry plantations. Raspberries may require some lime to bring the pH to the optimum 6.0 to 6.5 level.

Once the plantation is established, fertilizers (should they be needed) are usually put on in the spring by spreading between the rows and tilling in. Cow manure at 6 to 10 tons per acre is a typical addition. Mulching materials around the plants, such as sawdust or wood chips, require additional nitrogen. Most growers use ammonium nitrate at 2 pounds per 100 square feet.

In general, mulching is a good practice with raspberries. It should be 6 to 12 inches deep, and can be made up of almost any weed-free materials like straw, leaves, sawdust, wood chips, and grass clippings. These organic materials are fine preservers of moisture as well as good soil conditioners when plowed under.

E. Equipment needed:

The same tools are needed to fit the seedbed for the raspberry plantation as are needed for corn or potatoes. When planting rows on 6-foot centers, the cultivating can be done with a walk-behind rotary tiller (minimum of 6 horsepower). Tractor-mounted or drawn cultivators require rows on 10- to 12-foot centers. Otherwise all one needs are the usual hand tools, and a cane cutter — a sharp, V-shaped knife that is mounted on a 36-inch-long shaft. This back-saving tool can be easily fabricated in the farm shop.

Cane Cutter

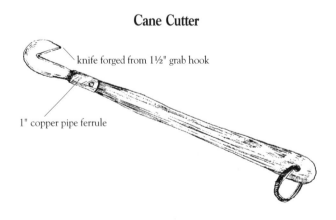

knife forged from 1½" grab hook

1" copper pipe ferrule

F. Field practices for raspberries:

1. **Seedbed preparation.** The raspberry seedbed is prepared in the same way as that for corn or potatoes, but the emphasis is on incorporating as much organic material into the soil as possible, and ridding the site of weeds (especially quack grass). Plow

down green manures whenever possible. Do not try to pulverize the soil to a fine powder, but break up clods and remove stones.

2. **Leveling.** Flood irrigation needs a minimal slope of 3 to 4 percent, and overhead irrigation requires careful leveling to avoid sags where water can stand.

3. **Pre-planting.**

a. **Choosing planting stock.** Only the best one-year-old, certified virus-free suckers should be used. Even with culled planting stock, the fussy grower should plan on discarding 10 to 20 percent of the new plants. It is "penny-wise and pound-foolish" to plant doubtful root systems. For an acre-sized plantation having 2- to 2½-foot in-row spacing of plants on 10-foot row-to-row centers, the farmer will have to buy about 2,200 plants.

b. **When to plant, and last-minute preparations.** As a rule, fall is a better time to plant than spring, but planting stock is harder to obtain at this time. If the new stock is planted in the fall in northern areas, it should be put in early enough for the plants to get established before winter. Green, unhardened canes will suffer winter dieback, and this could make a severe encroachment on early yields. Pacific Coast planting is best done when opportunities permit during the rainy season.

The roots of the suckers need to be kept moist until planting. If the time seems short (two or three days), cover them with a wet burlap sack, but if it looks like the planting time may be set back for a week or so, it is best to dig a trench and heel them in. Just before planting cull out the weakest-rooted plants, and then cut off the stem of each plant, leaving an 8-inch "handle."

c. **Row and hill systems.** Raspberries are planted in rows (hedges) or in hills. The hill system is usually employed only with black raspberries (fewer sucker canes leave room for air circulation). The hedgerow system is one where the rowed plants are permitted to produce suckers to fill in the intervals between plants. Spacing between these suckers is kept at about 6 to 8 inches by pruning. The width of the mature hedgerow at its base should be kept at 18 inches. In the hedgerow, the plants are supported by 12-gauge wire that is strung on 6-foot posts dug into the ground at 15-foot intervals. The posts should be secure (say, 2 feet underground), and for red and yellow raspberries the wire should be strung double

(one on each side of the post) at 2½ and 4 feet above the ground. Black and purple raspberries need only one double wire at about 3 feet above the ground. To keep from overly constricting the canes, notched spreaders are inserted where needed.

4. **Planting.** Soak the sucker's roots in muddy water for an hour before planting. A shovel is sufficient for most planting purposes — simply push the shovel 6 to 8 inches into the soil, tilt it forward, and then place the sucker behind it. Then lift the shovel out carefully, so as to avoid pulling the roots upward, and complete the planting by firming the soil with your heel. When properly placed, the leader buds (the ones just above the root system) should be 1½ to 2 inches below the soil surface. Black raspberries are planted slightly deeper than red ones. If time is short and the acreage large, open a furrow with a plow, and lay the suckers against the cut edge of the furrow. Kick or hoe dirt over the roots and firm it with your heel. Finish by plowing the remainder of the dirt back into the furrow. Always follow planting with irrigation.

Raspberry Training

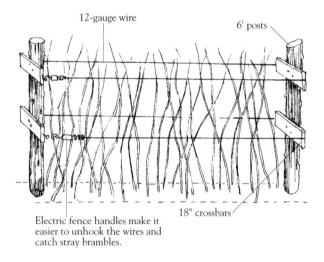

12-gauge wire   6' posts

Electric fence handles make it easier to unhook the wires and catch stray brambles.

18" crossbars

5. **Cultivation.** Cultivate only as often as necessary to control the weeds. This usually works out to once every seven to ten days. Never cultivate deeper than 4 inches so as to avoid pruning the root systems.

6. **Other spring work.** After planting, and after the first new shoots appear above the ground, prune off all the 8-inch "handles" that were left on for

planting. This old wood can act as a source of entry for diseases. No fertilizer is necessary the first year after planting, but, in years afterward, spring chores should include the incorporation of fertilizers and amendments between the rows. This is also the time to take new, green suckers to establish any new plantings the farmer may want.

Very early in the spring (starting the year after planting), the canes within the row are thinned to 6- to 8-inch spacings. This pruning is critical, for studies have shown that thicker canes produce more and larger berries. Usually the first suckers will prove to be the stronger and thicker ones. Finally, with red and yellow raspberries, the tips of the current year's bearing canes are topped (one-fourth the length the second year, and one-third the length thereafter). With black and purple raspberries, canes are topped in early summer, and the *laterals* (side branches) are pruned back in the early spring. This pruning is touchy, for the point at which the lateral is pruned varies depending upon the variety. In general, black raspberry laterals should be left 8 to 12 inches in length, and purple varieties should be 12 to 18 inches long. Leaving the laterals too long will produce more fruit this year, but this production will be at the expense of next year's crop.

7. Other summer work. Differences in yields prove the value of regular irrigating with raspberries. In drought periods, the plants should have an inch or more of water per week, during the fruiting period.

Early summer is the time for "pinching back" the terminal ends of new black and purple raspberry canes. Actually done with pruning shears, the pinching stops the end growth, and forces the lateral branches to grow in readiness for next year's production. Three or four pinchings of the plantation is necessary to catch all of these new canes. The optimal height for pinching back is 20 to 24 inches with black raspberries, and 30 to 40 inches with purple ones. The new laterals are allowed to grow all summer, lie dormant during the winter, and are pruned back the following spring. After harvest, all of the bearing canes are pruned out. This applies to red, yellow, black, and purple varieties. To prevent the spread of diseases, all prunings should be hauled away and burned, and the pruning process should be left to a time when there is no moisture on the leaves.

No fertilization should be done after the crop is harvested, as this will stimulate unwanted growth of green canes that will not harden off before winter.

8. Harvesting. Three-finger picking and placement instead of dropping into the basket are the secrets of unbruised fruit. Chest- or waist-strapped basket carriers hasten the picking, and can lessen the chances of bruising. Filled baskets can be stored in the shade while picking is going on, but they should be refrigerated at the earliest opportunity.

9. Other fall work. In areas where winterkill is a problem, the canes may have to be bent over and buried with soil or mulch. This covering is removed in the spring. Between-row cover crops are sometimes planted (to be plowed under in the spring), but this practice can sometimes lead to problems with volunteers.

10. Spraying. Most small-scale farmers do not spray their raspberries with chemical poisons, but it is wise to acquaint oneself with poisons to control cane borers and mosaics.

IV. ANALYSIS OF STRAWBERRIES AS A CASH CROP

A. General considerations:

Selecting a site, determining the size of an enterprise, and analyzing the suitability of soil and climate are much the same for strawberries as for raspberries. These berries also share the same market, and the methods of marketing are identical. Strawberries are different from raspberries in the following ways:

Strawberry

1. Strawberries bear only one year in most small-scale farm plantings (barring renovation). The first year is a planting and establishment year when all of the blossoms are removed. The second year is the bearing year, and then the bed is plowed under to be prepared for a new planting. In planning acreages to devote to a continuous cash crop of strawberries, the farmer who does not renovate his old beds must double (or better yet, triple) his bed space, keeping one in the establishment phase and the other harvestable. Growers' records indicate that one-year fruiting plans with strawberries are slightly more profitable than two- or three-year fruitings. If a third bed is figured into the rotation, one bed is kept in green manure cropping each year, thereby adding to the fertility and conditioning of the soil.

2. Strawberries require no physical control of their vertical growth. For commercial purposes they are best planted in rows that are trained into "spaced matted," 24-inch-wide beds. These beds are separated by a 30-inch lane (sometimes furrow) which is used for access and sometimes irrigation.

3. Strawberries can be picked faster than raspberries, therefore the farmer may wish to consider a larger kind of enterprise.

4. Yields average 3,000 to 4,000 quarts per acre, but may go as high as 10,000 quarts.

B. Varieties:

Choosing a variety that is best suited to the farmer's soils and climate and to his customers' preference is of primary importance. The greatest single determinant of varieties is climate. Farmers living in northeastern areas prefer Catskill, Robinson, Sparkle, and Midway because of their solid production records.

These varieties along with Dunlap are grown in the northern Great Plains region, but where the seasons are too severe, growers are forced to consider everbearers. Moving south from the Great Lakes region, growers show successes with Midway, Sparkle, Catskill, Tennessee Beauty, and Albritton (North Carolina). Blakemores and Dixieland are the varieties that have the most success in the Southeast, and Florida Ninety is the variety that is most adaptable to Florida's temperatures. Northwest, apropos of its name, is favored by growers in the coastal areas of Washington state and Oregon, and Shasta and Fern are California's favorite valley varieties.

Once the varieties best suited to his climate are determined, the farmer usually has the luxury of selecting other varieties that will extend his bearing season. This is particularly important to him when his is not a pick-your-own operation. The following extract from the USDA Farmer's Bulletin 1043 classifies common varieties by their ripening season:

Very Early	**Midseason**	**Late**
Earlidawn	Catskill	Albritton
Midland	Dunlap	Columbia
	Fairfax	Jerseybelle
Early	Fern (day-length	Siletz
Blakemore	neutral)	Sparkle
Dixieland	Honeoye	Tennessee
Florida Ninety	Jewel	Beauty
Howard 17	Marshall	
(Premier)	Midway	**Very Late**
Pocahontas	Shasta	Molalla
Surecrop		Redstar
	Late Midseason	Vesper
	Armore	
	Blomisan	
	Empire	
	Northwest	
	Robinson	

Farmers having pick-your-own operations have experienced management difficulties when they have more than one variety of strawberry in their patch. Customers have been known to trample across three rows of perfectly ripe berries to get at a section of unripe, later-maturing ones, despite row-picking assignments and "Keep Out" signs that are lettered in 3-foot-high block print. Farmers who have survived this kind of a contrary customer stampede now restrict their pick-your-own plantings to a single variety, and count on doing a larger volume of business in a shorter period of time.

C. Fertilizers and amendments:

Strawberries and raspberries require the same fertilizers and soil amendments. Only the time of application is different. Strawberry beds receive their biggest application of nutrients after harvesting, if the beds are being renovated.

D. Equipment needed:

Preparing the seedbed for strawberries requires the same equipment as is used for corn or potatoes. Once planted, the between-row tillage requires a walk-behind rotary tiller (6 horsepower is the minimum for an acre of berries). A power lawnmower is necessary if the bed is to be renovated.

E. Field practices for strawberries:

If he has enough room, rotating three beds is an ideal arrangement for the small-scale farmer. Having one bed seeded to green manure crops in its "off" year is an inexpensive way to fertilize and condition the soil while keeping weeds from encroaching on the patch. As with raspberries, the greatest enemy of the strawberry crop is quack grass. The farmer needs to be diligent in his pursuit of ways to control this weed, and fast-growing green manure crops and mulch are two of his most potent weapons.

If quack grass is the strawberry's worst enemy, drought runs a close second. Strawberries can be irrigated by flooding between the rows, or water can be applied with an overhead sprinkler system. Some growers use their sprinkler systems to apply liquid fertilizers. In either case, water should be given to strawberries at the rate of 1 to 2 inches each week during the spring and summer (less when rains have provided some moisture), and this practice should be employed both in the bearing year and in the year when the plants are sending out runners.

There is a correlation between the number and size of leaves on a strawberry plant, and its fruit yield. Nationwide statistics show that irrigated fields have 25 to 50 percent more yield, and that the plants are larger, hardier, and more vital.

1. First year.

a. **Pre-planting.** The practices here are the same as for raspberries.

b. **Planting.** Strawberries can be set in the ground in the same way as raspberries. The distances for planting are 18 to 24 inches in-row spacing, and 48 to 52 inches between-row spacing. They should be set 4 to 6 inches deep. (52" x 18" = 6,670 plants per acre; 48" x 18" = 7,260 plants p.a.; 48" x 24" = 5,445 plants p.a.) It is critical to set the plants so that the midpoint of the crown is at the soil surface level. Roots should be spread slightly, and they should point downward. Soil should be firmly packed around the roots, and every planting operation should be followed by a thorough irrigation (5- to 6-inch penetration). To assure a good planting, use only the best, most bushy-rooted plants. Discrimination here may result in throwing aside as much as one-third of

the planting stock, but the final yield results will justify being finicky.

Planting may be done in the spring (most common), or in the *early* fall of the preceding year. There is evidence to suggest that fall plantings produce more leaves and stolons, but planting stock is generally hard to come by at this time. A fall planting should be done early enough to allow the plants to get established and "harden off" before winter (plant by August in the North, September or October in the South).

c. **Post-planting, the first year.** The farmer's concerns during this period are weed control, blossom removal, daughter plant placement, and irrigation. Weeds between the rows can be controlled with the rotary tiller, but those between the plants must be pulled by hand (or hoed). When weeding in early summer, pinch off the blossoms. In the midsummer weeding, study and try to regulate the daughter plant placement. Ideal placement provides 6 inches between daughter plants, but one will learn that the neat diagrams found in most books do not take into account the burgeoning, all-at-once growth of these stolons. The farmer usually settles for establishing growing room for the daughter plants (pattern be hanged), and then he ruthlessly prunes off later plant runners.

All varieties except those grown in the South and along the West Coast stop putting out leaves in late August, September, or October (depending upon latitude), and concentrate on making fruit buds. Finding fruit buds at this time in the growing points of the crowns of the plants promises a good-to-average yield year, but also finding fruit buds on the leaf axils promises a bumper crop.

In the North, the farmer must cover his strawberry rows with a deep mulch after the ground has frozen to prevent crop loss due to frost heaving. The best mulch is straw because it has been threshed of its seeds and is less apt to bring weeds with it. This mulch insurance is most needed in those areas where the wind blows the snow off the patch and temperatures drop below 0°F. (–18°C.)

2. Second year.

a. **Spring.** The mulch should be removed when a yellowing of the leaves starts to occur and the leaves begin to grow. It is a good idea to fork the mulch over into the adjacent aisles until the last threat of frost has passed. If the temperature at *ground*

Runner Cutter

old putty knife

old broom handle

level drops below freezing, a 12-inch mulch will protect the plants to 26° to 27°F (–3°C.). The farmer who has no mulch can use large 4-foot-wide rolls of thick paper. If he is careful the paper will last for several seasons. About the best frost control is an overhead irrigation system. If the water is turned on at 32°F. (0°C.) and kept on until all the ice on the blossoms has melted and there is a layer of water between the ice and the leaf surface, damage can be kept to a minimum. Caring for strawberries for an entire year and then losing them to frost a month before harvest is lousy economics. The strawberry farmer *must* provide for frost protection.

The temptation for beginners is to fertilize their bed in the spring of the bearing year. In most cases this is neither necessary nor desirable. If the soil was fertilized properly in the pre-planting phase, and it has plenty of organic matter, it should not need fertilizing again during the growing year.

b. Harvest. Thirty days after blossoming the fruit is ready to harvest. The bed should be picked early in the morning every two to four days, and every ripe berry should be taken at each picking. If birds are a severe problem (this usually occurs after the berries begin to turn red), netting should be laid over the rows for the entire harvesting season. Losses to birds can be severe in some areas, and the farmer must figure the cost of netting into his overhead.

c. Post-harvest. The day after the last picking, the farmer should start renovating the old bed, or plow the plants under preparatory to seeding in a green manure crop. Renovation entails:

1. Weeding the bed thoroughly.
2. Mowing the old strawberry plants to a height of 1 or 2 inches. One should be careful to avoid mowing the crowns off, as they are necessary for the plant's continuing growth.
3. Rotary-tilling the aisles to a 6-inch depth. This tilling *should* encroach on the bed sides, thereby cutting the width of the beds back to about 12 inches.
4. Weeding out (roguing) weak or diseased plants.
5. Fertilizing the aisles with cow manure at the rate of 10 to 12 tons per acre or a complete fertilizer (15–15–15) at 750 to 1,000 pounds per acre. Rotary-till the fertilizer into the top 4 to 6 inches of soil.
6. Praying for rain or irrigating until the soil is moist to a 6-inch depth.
7. Treating the bed as a new one thereafter. (See the practices for the first year.)

Many renovated beds will return up to a third less yield the following year, and their economic viability is questionable for the small-scale farmer. In Vermont, for example, a field survey showed that half the growers renovated, and half fruited their beds for one year only. Each farmer must make his own decision, but if there is any question about the thriftiness or health of the plants in the bed, the bed should be turned under and prepared for new plants.

CHAPTER 7

The Whole Farm

Cash crops are not the alpha and omega of the small-scale farm. No one can properly manage a farm without devoting a part of his attention to its supportive components. Farmstead road, water, woodlot, shop, and fencing are necessary and integral parts of the whole farm, and they should be planned, developed, and built on a long-range master-plan basis.

ROADS

Access roads to the farm itself, and the farm's internal road system need planning with an eye to the other components of the farm. As a general rule, roads should be kept to a minimum in productive fields to avoid compaction and the deterioration of water-control constructions, but when access or traverse is essential, it is best to use field-edge headlands for that purpose.

Permanent roads, such as those leading to the farmhouse or barn, are worth the expense of putting them in right. Tarred roads are still the exception in farm country, and most small-scale farmers will content themselves with an evenly-graded, well-drained road with an aggregate topping. Permanent roads should be at least 20 feet wide, and graveled, or topped to a depth of 12 to 18 inches with whatever porous, long-wearing material is locally available. The road should be crowned (5 to 6 inches) in the center, and the shoulder should slope to an evenly graded drainage ditch on the uphill side. On lengthy, cut-and-fill stretches, the uphill ditch should be drained with culverts placed at 300- to 500-foot intervals, and the uphill bank and culvert outlet should be lined with stone riprap.

Roads that see only seasonal use should be graded gently to prevent erosive rutting, and then seeded to a permanent sod. Where sod is impractical, such as on roads running through dense woodlots, drainage diversion cuts should be made across the road to keep them from washing out.

Building roads to cross watercourses or perennially wet areas around the farm should be avoided whenever possible. Bridges and elevated roadbeds are expensive to build and maintain, and, unless there are no other options, most farmers go around or make do with cobbled crossings that are negotiable at times when they are needed.

FARM WATER

The presence of water has its drawbacks in building roads, but without it one has no business farming. The farmer has two distinct water needs — first, for his crops, and, second, potable water for himself and any livestock. These needs are measurable, and determining the quantity and the sources of supply are prerequisite to farming. Farms located in regions of normally dependable rainfall still must possess or have access to reliable potable water sources. In regions having supplementary irrigation, water sources should be regarded as a bonus and a marketable asset.

In this connection, it is interesting to look at the water-farm relationship through the eyes of a professional farm appraiser. In a recent manual for farm appraisers, fully one-fourth of the instructive text deals with the availability, quality, and cost of water in determining the worth of a farm.

A farm buyer needs to look at any farm with the dispassionate eye of an appraiser, but when the farm depends on irrigation water he should also become a detective. Before buying he should investigate:

1. The appropriate doctrine of access (riparian or prior appropriation) that applies to his prospective farm (see page 43). Water rights are usually shown by water right decrees; state certificates; irrigation district locations; contracts with the U.S. Bureau of Reclamation, the Secretary of the Interior, or a public water utility; stock shares in a company; or simply a riparian location.

2. Any engineering reports made by federal, state, or private engineers concerning the farm's water supply or drainage.

3. Any remaining debt liability connected with the farm's participation in a water improvement

Measuring Small Quantities of Water

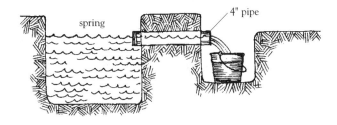

Record to the second the time it takes a bucket of known capacity to fill. Then convert that figure to gallons per minute.

district or other irrigation enterprise.

4. Historical amounts and costs of the farm's water supply. Delivery and cost records are normally available if the water is obtained from an organized conservation, irrigation, or special improvement district. Should the source of supply come from streams, rivers, or groundwater pumping stations, the investigation should include meteorological records of precipitation, and any governmental or private reports concerning the local water table. If pumping plants are involved, the buyer should also know the costs of the pumping operation (energy, depreciation, labor, etc.), and he should determine the pump's efficiency compared to the recorded value of the crops produced.

Whether researching a farm or operating one on a day-to-day basis, the farmer needs to know how water flows are measured. Smaller water sources like springs, brooks, or wells produce flows that are usually measured in gallons per minute (g.p.m.). Larger sources like rivers, lakes, or ponds have flows that can also be expressed in g.p.m., but are more commonly measured in cubic feet (cu. ft.), acre-inches (ac. in.), or acre-feet (ac. ft.). Flow measurements are expressed in time intervals, as cubic-feet-per-second, acre-inches-per-hour, and acre-feet-per-24 hours. To confuse matters further, in some western states, flows are often measured in an historic oddity called "miner's inches," and the definition of a miner's inch will vary from state to state.[1]

---

[1]In northern California (also Arizona, Montana, and Oregon), one cu.ft./ sec. equals 40.0 miner's inches, and in southern California (also Idaho, Nevada, New Mexico, North Dakota, South Dakota, and Nebraska) it equals 50.0 miner's inches. Colorado differs from everyone by claiming that one cu.ft./sec. equals 38.4 miner's inches.

Gravity Flow Water Systems

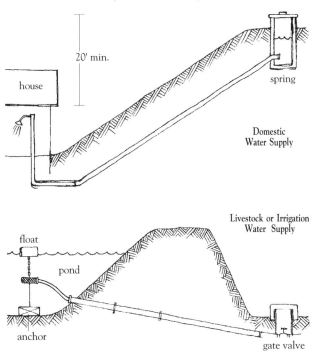

Gallons and cubic feet are self-explanatory units; an acre-foot of water is an acre area that is one foot deep (43,560 cubic feet); and an acre-inch is one-twelfth of that amount. The term *miner's inch* derives from the gold rush era in California when miners were allotted timed amounts of water that was metered through a barely covered, 1-inch hole. They used the precious water to sluice the gold, and to wash the ore from bar deposits with hydraulic nozzles.

Though less romantic, irrigation water still produces "gold" today in the form of farm crops. Despite our improved technical capacity to move water, we still rely, for the most part, on gravity for getting water to our irrigation systems and to some of our domestic water supplies. The fact that gravity systems have comparatively few operating expenses accounts for their continuing popularity, and farmers turn to electric or fossil-fuel powered pumps only as a second choice.

But gravity's effect on water is a problem as well as a blessing. Water that is drawn by gravity to the sea is lost to farm crops, and it carries with it valuable topsoil and soil nutrients. The faster the rate of runoff, the greater the erosion. It is therefore the farmer's task to slow the runoff to the point where his crop roots can take the maximum amount, and where the least amount of his soil will be washed away. This is called water management.

Aside from using farm practices that prevent his soil from eroding, the farmer's most gratifying water management accomplishment is building himself a farm pond. Having water in storage is like money in the bank, and the practical advantages are augmented by the pond's recreational potential. The most common practical uses for a farm pond are those of livestock watering and irrigation, but they are also used for fire control (installation of a dry hydrant in cold climates), and for hydroelectric power generation (providing the inflow is of sufficient quantity and reliable).

Nearly 2.5 million ponds have been built in the United States. They vary in size from several acres to pothole-sized catch-basins, and they are common to all regions of the country. The size and location of a farm pond are limited by economic considerations, the uses to which the water is put, the presence and reliability of water supply, the type of soil at the site, and topography.

Few ponds are dug by hand today, and not many farmers have the equipment to do the job. Therefore the average farmer hires outside earth-moving contractors to do the work. There are two basic kinds of ponds (embankment and excavated), and both require the movement of large quantities of earth. An embankment pond is usually built by putting a dam across a stream or other watercourse, and it is typically sited in depressions that will permit a minimum water storage depth of 6 feet (deeper ponds for more arid areas). Excavated ponds are those made by excavating a pit.

Considerable planning should precede the ground-breaking for a pond. The farmer should know, for example, how much water he will need. If his needs are to water livestock, he should estimate his gallonage needs per day (15 gallons per head for cattle, horses, and dairy cows; 4 per head for hogs; 2 per head for sheep), and then he should balance these needs off against the projected reliability of the water source.

Irrigation storage requirements are greater than any others discussed here, and the pond's planned capacity has to be large enough to meet the farmer's crop needs for an entire growing season, plus allowances for evaporation, seepage, and other losses.

Fire-Fighting Dry Hydrant for Farm Pond

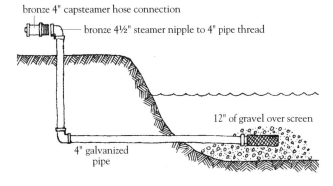

bronze 4" capsteamer hose connection

bronze 4½" steamer nipple to 4" pipe thread

12" of gravel over screen

4" galvanized pipe

Irrigating from a pond entails a lot of water storage. For example, 3-inch flooding of an acre-sized orchard would use up 81,462 gallons of water. Because of the severe fluctuations, ponds used for irrigation should not be used for recreation or wildlife habitat purposes.

Ponds that are built on streams or rivers are less susceptible to problems of drought cycles than those depending upon surface runoff. In drier areas of the Midwest, for example, a ¾-acre pond averaging 6 to 7 feet in depth might require a drainage area of 200 acres. By contrast, a drainage area of this same size in New England might require prohibitively expensive overflow structures for excess runoff during storms. Of course, other drainage factors such as ground cover, steepness of slope, and the soil's ability to absorb water would enter into calculations of how large a farmer might build his pond or what kind of a spillway he would need.

There is no substitute for the farmer finding out for himself the local runoff rates and then assessing their implications for his pond site before he begins construction. He can determine these rates by contacting his nearest Agricultural Extension agent, or Soil Conservation Service district agent. If neither of these resources is available, he should write to the U.S. Government Printing Office, Washington, D.C. 20402 for a copy of the U.S. Weather Bureau Rainfall Frequency Atlas, Technical Paper 40.

Suitability of a pond site depends on several other factors besides peak drainage area discharge. Most important, a pond's soils must be able to hold the water in the reservoir area. To determine what kinds of soils are present at a pond site, one should take a series of core samples from the impoundment area,

and from the area of the proposed dam or dike. Clay and silty clay are ideal embankment pond sites, and sandy clays are usually satisfactory. Areas of coarse sands, gravel, and limestone are too porous, and should be avoided. There are materials one can import to seal leaky or potentially leaky ponds, but they are expensive and should be considered only as a last alternative.

Since it is assumed that all mucks, peats, and organic material will be stripped from the pond site, the pond builder should concern himself with the contents of the core samples below such layers. Trouble-free earth-fill dams need at least 20 percent (by weight) clay soil, and it makes economic sense to take this material from the excavations. Therefore one should submit these samples to a bottle test: Add water to fill a quart jar that is one-third full of the soil to be sampled (remove the larger stones first); shake and allow the contents to settle; the finer material (clays) will be the last to settle; measure the depth of sand, silt, and clay with a ruler to determine proportion.

Once the site and size of the pond are decided upon, the farmer should begin planning his dike and dam locations. To do this he should make a simple survey with a level preparatory to making a contour map. Again, help with this survey can be obtained from SCS or local agricultural agents, but the farmer could manage to survey the smaller pond sites himself.

Do-it-yourselfers need a carpenter's level, a homemade survey rod (a 10-foot, or so, pole that is marked in inches), a 50-foot measuring tape, and an assistant. Using the tape, grid the pond site at 25- or 50-foot intervals, and place survey markers at the interstices. Placement of these markers needs only "eyeball" accuracy, but it helps to use two or more of the markers that have already been placed to line up the ones that follow.

Before taking readings of the marker sites, establish a reference point (bench mark) that is likely to remain undisturbed during the earth-moving process. This bench mark will provide a point to measure the progress of excavation. To take readings of the bench mark and the marker sites, place the carpenter's level on a wooden staff or piece of lumber, and sight across the level to the survey rod which is being held by an assistant. The assistant can help the surveyor with the more distant readings, indicating the level with a brightly colored indicator on the face of the survey rod. A mirror held above the bubble on the level will allow the surveyor to sight the rod at the same time that he is trueing the level.

The next step in contour mapping is to transfer the field readings to a piece of graph paper, and then to draw in contour lines (points of equal elevation) that are of an even interval. One-foot intervals are usually used for level terrain, and 2-foot intervals for steeper slopes. Obvious dike or dam locations will emerge from the map, and from them one is able to plot the prospective water line and the spillway height. If the prospective pond is being planned for an area where springs are likely to be discovered, it is a good idea to provide for a possible enlargement of the site.

Should an estimate of the cubic yardage of earth to be moved be needed (some contractors base their estimates on this factor), a rough one can be arrived at by using the field readings to make a graphed profile (usually on the centerline) of the pond site. By superimposing the desired final line of the pond bottom below the present earth line on the graph paper, one can make a fair estimate of the quantity of earth that will have to be moved.

Pond planning does not end with location. Dams and dikes still have to be designed for their particular locations. The top width of the dam should be 8 to 14 feet, depending upon its height, and the earth fill is always more gradually sloped on the pond side than on the downstream side. The proportion is usually 3 feet horizontal to 1 foot vertical upstream, and 2:1 downstream.

Larger dams or dikes are constructed with two control factors built into them: an overflow pipe (trickle-tube) and a spillway. The spillway is intended to pass excess storm runoff around the dam rather than over it, thus avoiding washouts, and the overflow pipe is designed to handle the normal discharge flows from storms, springs, snowmelt, or seepage. As the first line of defense of the dam, the overflow inlet is placed so that it is 12 inches or more below the spillway level, and its flow capacity (diameter of opening) is sized to cope with the drainage area's runoff rates at maximum historical levels.

Calculating the size of overflow pipe, the design of the spillway, and the pond's drainage requirements for an earth-fill dam are not necessarily the exclusive

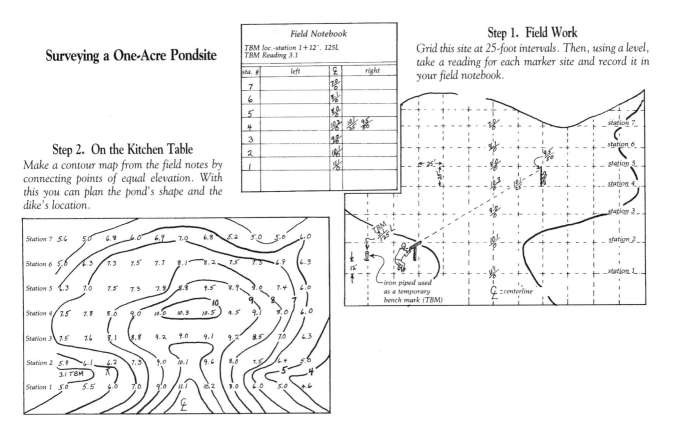

Surveying a One-Acre Pondsite

Step 2. On the Kitchen Table

Make a contour map from the field notes by connecting points of equal elevation. With this you can plan the pond's shape and the dike's location.

Step 1. Field Work

Grid this site at 25-foot intervals. Then, using a level, take a reading for each marker site and record it in your field notebook.

province of professional engineers, though their services are available to the pond builder through the Soil Conservation Service. If the farmer does not make these calculations himself, he should, at least, understand how they are arrived at. To prepare himself, he should first research the local codes having to do with pond construction, and then write to the U.S. Government Printing Office, Washington, D.C. 20402 for the instructive pamphlet, *Ponds — Planning, Design, Construction,* Agricultural Handbook 590.

After all the advance planning, the actual construction of the pond seems anticlimactic. The first step is to clear and dispose of vegetative matter like trees, brush, and roots from the pond area, beginning 15 feet outside the pond's planned water line. Layers of peats, mucks, or other organic matter should be removed before beginning to dig the cutoff trench. The latter is simply a drainage route that is intended to keep water from accumulating in the working area. Topsoil should be set aside.

Stakes should be placed to guide the equipment operator in the construction of the dam or dikes. A line of them along the centerline of the dam may be enough if the pond is small and the operator experienced. But more often than not, the outer perimeters of the dam (upstream and downstream) should also be marked. It is also wise to stake the spillway and the overflow pipe locations. Anti-seepage collar positions on the overflow pipe should also be clearly marked.

At this point the earth-fill for the dam is begun. Most dams are constructed from the material that is excavated from the impoundment area, but fill may have to be imported from a nearby "borrow" pit. Brought to and laid down in the dam in 8- to 10-inch layers, the fill is compacted between layers. This is usually done by the earthmoving equipment, but larger dams or difficult soils may require special rollers. During the course of filling in the dam, the overflow pipe is put into place, and material around the anti-seep collars is carefully tamped to prevent later weakness developing as the soil settles.

If the cutoff trench is made through the dam line itself, it should be pumped dry and carefully backfilled and tamped. Lastly, the earth spillway is fashioned to the predetermined specifications, making sure that the bottom of it is level from side to side and at least 12 inches above the top of the upstream end of the overflow pipe, but below the top of the dam.

When fill quality varies, the most impervious should be placed in the center of the upstream side of the dam. More importantly, all fill should be carefully compacted whether it is of coarse or fine consistency. The moisture content of the fill will largely determine how successful the compaction will be; therefore, before beginning, the pond-builder should hand-mold some of the fill and roll it out between his hands.

If the fill from the impoundment area or the borrow pit met the requirements of the earlier bottle test, it is ready for compaction if it molds as it rolls, but shows a tendency to crumble. Should the sample not mold, it is too dry (in this case, sprinkle it or wait for rain), but if it molds and rolls to the thickness of a pencil, it is too wet.

Once the dam is completed, it should be covered with the topsoil saved (if any), then immediately seeded to sod-type grasses. Ironic as it may seem, pond construction is usually done during the driest time of the year, and the new seeding may need supplementary irrigation. In any case, it is always a good practice to cover the seeded area with a thin mulch of straw, hay, or fodder. The best kind of seed to use depends upon the climate and kinds of soils, but most mixtures include some varieties of fast-growing perennial grasses and legumes.

THE WOODLOT

Like a pond, a woodlot is a plus for any farm. Farm appraisers list this plus as pulp, sawlogs, or under the headings of products like maple syrup or turpentine. The knowledgeable farmer knows that harvestable crops are not the only values of his woodlot. He knows that trees are unsurpassed as erosion controllers, and though he is not usually inclined to poetic description, he has a working, day-to-day appreciation of the beauty of his woodlot, an appreciation that the poet seldom shares.

Woodlots come in all sizes and shapes. They vary from windbreak strips between fields to solid chunks of established forest. They serve as wildlife habitats, and as a potential source of forage for some domestic livestock. But, historically, the farm woodlot has been a renewable source of fuel and farmstead building material. The past three decades have seen this use diminish, but the pendulum has reversed its swing

Usually locally administered, the FIP can be tapped for help in planning forest management, purchase of nursery stock, and maintenance of the farm woodlot. A call to the county or state forester will bring him to the farm, where he will map the farm and make recommendations concerning the woodlot that will dovetail with the farmer's overall farm plan. If funds are budgeted (funds are administered by the Agricultural Stabilization and Conservation Service in most states) and available, and if the particular situation calls for it, he may also help the farmer to apply for financial aid in thinning a potentially productive stand (the thinnings or culls are the farmer's to sell or use for firewood), or he may recommend aid in installing or improving existing woodlot access roads.

Policy varies from state to state, but in many instances FIP funds are used to establish and maintain reforestation nursery stock that the farmer can buy at bargain rates. Red pine, Jeffrey pine, Scotch pine, white spruce, and walnuts are just a few of the seedling stocks that this program makes available.

with recent energy shortages and the inflated prices for raw building materials. Small-scale farmers, particularly those who are just getting started, will find that a healthy woodlot is ideally suited to their particular needs.

Because most trees take so long to grow, a harvestable crop of pulp, sawlogs, or maple syrup may not occur during the lifespan of the current owner (25 to 30 years are usually required for pulp, and 60 to 70 years for sawlogs). Thus, planting acreage to trees, be they hardwoods, softwoods, or orchards, is often an act of stewardship to the land and to those who will later reap the benefits. Altruism has its limitations for the small-scale farmer (particularly in the region of his pocketbook), and he may well decide that he cannot afford a new planting.

But a sense of dedication is not the only reason for the small farmer to consider reforesting a section of his farm. He may discover that not planting is "penny-wise and pound-foolish." Erosion is a juggernaut force that costs millions of dollars worth of topsoil each year, costs that can be curtailed (if not stopped entirely) by the planting of trees. To "sweeten the pot" the federal government has funded a Forest Improvement Program (FIP) that is designed to help the independent woodlot owner establish and maintain forest woodlots.

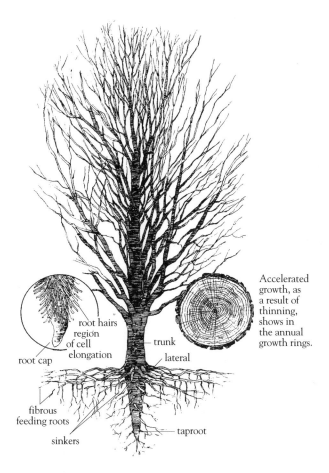

root hairs region of cell elongation

root cap

trunk

lateral

Accelerated growth, as a result of thinning, shows in the annual growth rings.

fibrous feeding roots

sinkers

taproot

This stock is mostly two years old, and can be had (at this writing) for as little as $50 to $60 per thousand.

In small-scale woodlots, small nursery stock is usually planted, although it can be grown from seeds. Though seed is less expensive to buy, the labor factor required for thinning, the growing time lost in bringing seeded plants to comparative size, and the time lost to reseeding "non-catch" areas (birds and rodents take a heavy toll of seed), make starting with nursery plants the better alternative. In either case, little equipment is needed to plant a new small forest — a mattock or shovel will do nicely for plants, and a mechanical, spinner-type seeder does a satisfactory job with seeds.

Not all new plantings of cash crop trees are beyond a farmer's lifetime realization. Christmas trees are a perennially marketable item, and using the relatively new methods of harvesting, such as cutting the tree so that the lower branches are left to propagate new stock, one can avoid a yearly planting expense. Fir, spruce, and pine are hardy, fast-growing trees that can be used for this purpose, and, if properly pruned, thinned, replaced, and "weeded," they require virtually no other care. They can be harvested on a rotating three-, four-, or five-year basis, and each root stalk can provide as many as five or six cuttings.

While new seeding is a problem for many farmers, most already have an established woodlot, one which presents quite different problems. Typically, an established woodlot has been cut over once, and the

second growth is unevenly aged, mixed in species, and too dense to allow healthy growth. This neglect can usually be traced back to the abandonment of the woodlot as a source of heating fuel.

Stewardship needs to be reasserted with these stands, and the first step is to assess what trees of those present are best suited to the soil, climate, and local market. Solid coniferous stands are common to western woodlots, but eastern ones often have softwood and hardwood mixed. Mixed stands are not necessarily incompatible. In fact, all too little is known about the benefits of some combinations of trees. Beeches and dogwoods, for example, complement maples, as they seem to manufacture a mineral (calcium) much needed by the maples. Certainly the farmer will find needs around his farm for building materials of both hardwood and softwood.

Once the desirable varieties are determined, the woodlot should be thinned and weeded to provide the best possible growing medium for the selected trees. Different trees require different spacings. Christmas trees, for example, can be spaced as close together as 5 feet, but maple trees intended for sugaring should be 20 or 30 feet apart. Spacing varies both with the type and with the age of the tree. Young, fast-growing pines, for example, that are intended for sawlogs may be spaced 10 or 11 feet apart, and trees intended for cordwood or fence posts may be thinned to 8 feet apart.

All too often, weeding the woodlot is assumed to mean the removal of *all* fast-growing, low-market-value trees. There are cases where these trees should be left. For example, fast-growing willows, alders, or cottonwoods may be essential to retain soil on erosion-prone areas like streambeds or river banks. In the Northwest the red alder, once considered a weed tree, has been found to be a desirable companion planting with Douglas fir because of its nitrogen-fixing ability and its capacity to help the firs resist root rot (the Northwest's greatest single forest disease). Clearly, common sense and some local research are essential before attempting a weeding project.

As time passes and the trees grow, their crowns begin to impinge upon one another, and they need to be thinned further. It is said that the definitive characteristic of a good forester is a perpetual crick in the neck acquired from a constant study of tree crowns. Crown crowding produces unhealthy

Planting Woodlot Trees

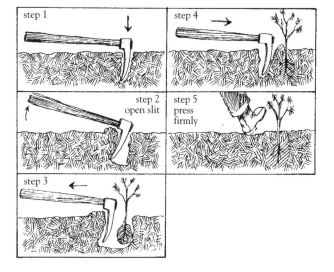

step 1

step 2 open slit

step 3

step 4

step 5 press firmly

Woodlot Thinning

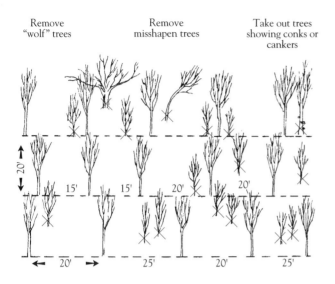

| Remove "wolf" trees | Remove misshapen trees | Take out trees showing conks or cankers |

Trees need space to fully utilize available nutrients. The optimal spacing for most trees is 20 feet or roughly eight paces.

conditions that result in stunted and unthrifty trees. Whenever crowns are crowded or are less than one-third of the total height of the tree, some thinning is indicated. Here, experienced judgment is indispensable in choosing which trees to remove, and the beginner should call in outside consulting help, such as the county, state, or federal forester.

Small woodlot owners who rely on their forests for firewood should regard the thinning process as both a harvesting and a conditioning for future harvesting. Because many woodlots are small and usually intended as a constant source of fuel and building materials for the farm, farmers are wise to forget clear-cutting as a woodlot harvesting concept.

Thinning decisions can be based on several factors. One might choose to cut on the basis of tree girth or height, or the selection might be predicated upon eliminating "undesirable" trees that inhibit growth of the major species. If the woodlot is of the self-regenerative type, the farmer must maintain a constant growth of thrifty young saplings and seedlings. To give the young saplings a chance, whole overtopping trees or, when possible, "wolf" trees (big, overripe trees that have spread to overshadow more than their share of the forest) may have to be removed.

Diseases in the woodlot are a considerable factor in woodlot thinning. Heart rot, for example, is a fungal disease that is the greatest single cause of woodlot losses, and, once infected, the trees are unredeemable and should be cut down. This disease is spread by wind- and animal-borne spores. It enters a tree at natural or man-caused wounds where heartwood is exposed, and it shows its presence by conks or fruiting bodies on the outer bark of the tree. Heart-rotted trees are seldom good for anything but firewood.

Canker-diseased trees may need to be removed, depending upon the extent or type of fungal canker. Cankers can show themselves as large, bleeding bulges in the trunk or limbs of the tree, or in tiny, orange-colored rust blisters on the bark. Blister rusts are restricted to five-needled pines, and detection of the blisters warrants immediate eradication of all nearby intermediate host plants like gooseberry or currant (in the case of southern pine fusiform rust, the intermediate host plants are oaks). Removal of these host plants will stop the course of the disease, but rust blisters will continue to crop up on the branches of younger trees for three years after the host plant eradication. These trees can be saved by pruning out the infected areas.

Other kinds of rust do not rely on intermediate host plants. They rely on direct contact for their spread, and control can be established by removing infected trees. Western gall rust is such a disease, and it claims millions of board feet of Jeffrey, ponderosa, lodgepole, and Digger pines in the West each year. Although it is not economically feasible to try to save trees that are already infected, the wood from these diseased trees is usable, provided the trees are taken early enough.

Disease control can be furthered in the thinning process by cutting to encourage the growth of disease-resistant species of trees. Marking for selective cutting of this sort is, generally speaking, the province of the experienced forester, and the beginner should ask for the help of the nearest county, state, or federal forester. Operating from a regional picture, the forester can make recommendations from a perspective of a regional "woodlot" normally unavailable to the individual woodlot owner.

Disease and fire are two woodlot hazards that know no woodlot boundaries, and the small-scale woodlot owner should expect to cooperate in regional disease and fire-control programs.

Periodic thinning and selective cutting will keep

The Art of Felling

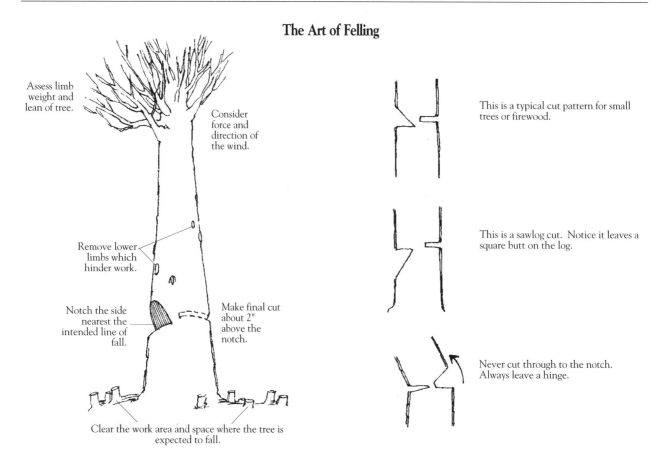

Assess limb weight and lean of tree.

Consider force and direction of the wind.

Remove lower limbs which hinder work.

Notch the side nearest the intended line of fall.

Make final cut about 2" above the notch.

Clear the work area and space where the tree is expected to fall.

This is a typical cut pattern for small trees or firewood.

This is a sawlog cut. Notice it leaves a square butt on the log.

Never cut through to the notch. Always leave a hinge.

a woodlot healthy and vital, and a healthy woodlot, though it be as small as 5 or 10 acres, can be expected to yield a surplus crop that can be sold. In most cases, the farmer should do this harvesting himself. The reasons for this are that the small scale of the woodlot and the fact that it is to be selectively cut will deter the interest of most logger operators. More importantly, stumpage contracts, no matter how tightly written, relinquish a measure of control that the farmer-steward cannot (and should not) afford.

But if the farmer has no other alternative, he should do some walking and some researching before entering into a contract with a logger operator. He should first have an idea of the best markets for his wood. Among the possibilities are sawlog timber, pulp, fence posts, veneer, railroad ties, telephone poles, cooperage, tool handles, artisan woodenware, charcoal, and firewood. With these possibilities in mind, he should walk his woodlot and mark those trees to be taken with two paint marks — one 4 to 6 feet above the ground, and the other below the stumpline. The lower one is to indicate that the cut was authorized.

While marking the trees, he should figure out how much wood is involved in the sale. Diameter measurements of the standing trees should be done at breast height (4½ feet), and the merchantable height (depending upon the anticipated use) should be estimated. Tables are available from government foresters for converting these figures into board feet or cord quantities. Because individual species vary greatly in their usable wood content, only tables that pertain to the specific species should be relied on. From these estimates, the farmer should subtract those sections of trees that are not suitable for the intended use, due to deformities, crookedness, rot, cracks, lightning scars, etc.

Only after the quantity of saleable wood is known should the farmer begin shopping around for a reputable logger operator. The surest way to find this person is to talk to neighbors who have had the same kind of work done. It is a good policy to visit the logged woodlot to see how the operator cleaned up after himself. When possible, it is also a good idea to visit a site where the logger operator is working and observe what care he exercises.

Tools for the Woodlot

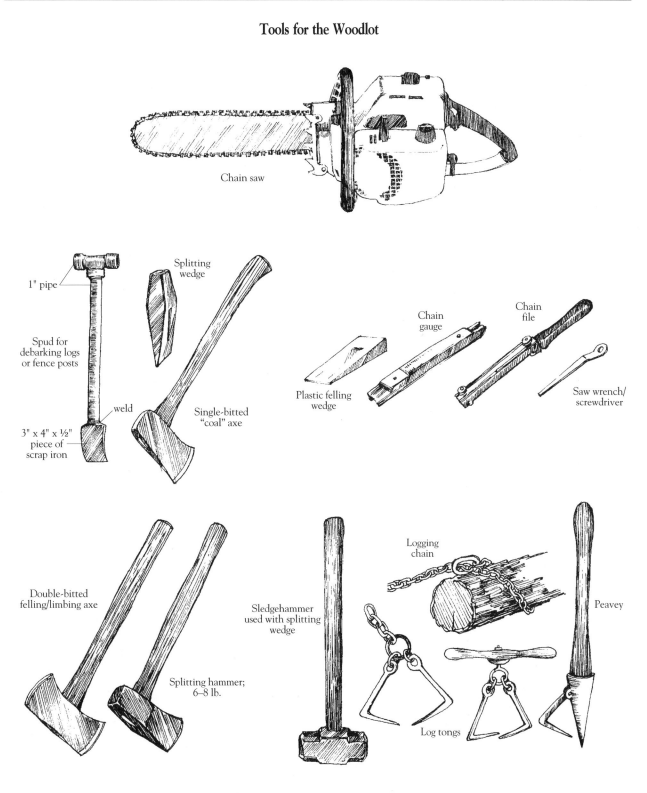

Chain saw

1" pipe

Spud for debarking logs or fence posts

weld

3" x 4" x ½" piece of scrap iron

Splitting wedge

Single-bitted "coal" axe

Plastic felling wedge

Chain gauge

Chain file

Saw wrench/screwdriver

Double-bitted felling/limbing axe

Splitting hammer; 6–8 lb.

Sledgehammer used with splitting wedge

Logging chain

Log tongs

Peavey

Log-scaling rule

Before entering into an actual price negotiation, the farmer should learn current prices being paid for stumpage. These prices can vary wildly within a given local area. The variables are timing, demand, quality of wood, and the business acumen of the parties involved. Dickering is not a lost art, as the farmer will discover when he finally approaches the potential contractor.

A handshake still seals the deal in many parts of the country, but in the long run more good feeling is preserved than lost by a legally witnessed timber sale contract. Sample contracts can usually be obtained from local governmental foresters.

Much of the research involved in the sale of stumpage should also be done by the farmer who harvests his own woodlot crop. He should know how much wood he has, and what the market is at the time he plans to harvest. The advantages to harvesting one's own crop are much the same as with any other farm product. The profit that accrues from labor, equipment, and management are the farmer's own, and unlike the logger operator, he is not as subject to the rise and fall of prices for his product. If the prices are down he knows that he can wait until they rise. His trees will continue to grow, and a season or so longer in the harvesting makes no particular difference to him. This latitude extends to the time that he actually does the harvesting. In most cases, he can arrange to do the work when it does not interfere with his other farm work.

Perhaps the greatest single advantage to harvesting one's own woodlot crop is the care with which the job is done. There is nothing like ownership to insure that young trees are left unharmed, and that maximum precautions are taken against fire while the woodlot is being worked.

How far one farmer can go with the logging operation depends upon what market his crop is aimed at, and how much equipment he has. If, for example, he is cutting firewood, he probably can manage the job with his farm equipment. But if his crop requires handling and transporting whole logs, he may have to hire the services of a trucker or skidder or both. Arrangements for these services and a market price should be established before the farmer cuts a single tree. It is a good practice, for example, to have a trucker look over the yarding site beforehand to insure that the location is, from the trucker's point of view, accessible.

Minimum tools for use in the farm woodlot include a chain saw and accoutrements, an axe (single-bitted ones are safer), a peavey, a splitting wedge (and/or a splitting maul), a felling wedge (plastic or wood for use with chain saw), a sledgehammer, a log-scaling rule, and a hard hat. If fence posts are to be harvested and debarked, a spud may also be required.

The chain saw has, because of its speed, revolutionized woodlot work. One mounted with a 16-inch bar and having a 50 to 60 cc. displacement is adequate for almost all woodlot practices. With it a good worker can fell, trim, and buck up as much as 2 or 3 cords of firewood per day. Of course, this excludes the most time-consuming jobs of brushing, splitting, hauling, and stacking.

But as the chain saw's speed is a blessing, so too is it a hazard. The speed and ease with which this tool cuts through a tree imparts a false sense of security that has to be guarded against. The normal respect one displays for edged tools should be doubled when the tools are powered, and redoubled when their cutting edges are unguarded. Felling trees with a chain saw is an occupation that requires concentration, an uncommon lot of common sense, and a diligent application of oneself to the acquisition of woods-lore. Like influenza, woods-lore is communicable and easy to acquire if one associates with the right people. Perhaps the easiest way for a beginner to pick up this knowledge is to first acknowledge his inexperience to a woods-wise neighbor, and then volunteer his help in return for the working experience he will gain.

FENCING

Because most farm fences involve wood in their construction, they are inextricably tied to the farm woodlot. Farm fences serve two purposes. The first is to keep livestock in, and the second is to keep pests out. Examples of the latter are fences to keep deer out of orchards, fences to keep domestic dogs away from sheep or goats, and fences around berry crops to halt the depredations of birds and rodents.

The problem with wire and wood farm fencing, even that which is well constructed, is that it rots or rusts, and is a constant source of expense and time spent on repair. Wood, no matter how resistant, eventually succumbs to fungal rot, and wire, galvanized or not, eventually oxidizes to lines of rust.

European farmers discovered years ago that the

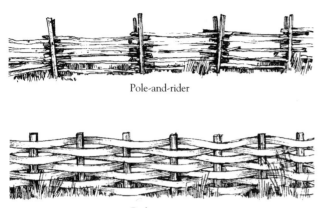

Pole-and-rider

Basket-weave

most trouble-free and cheap fence they could build was one of stone. Stone fences come in two groups: dry-laid and masonry-assembled. Except for retaining walls, dry-laid fences require larger quantities of stone (because of the requisite thicknesses) than do masonry-assembled ones, but the latter are more costly because of the masonry ingredients.

Building a stone wall without masonry is a long-term project that requires planning. Like any other stone structure, it is not portable, and once it is placed it is there to stay. Stone fencing makes sense on the small-scale, labor-intensive farm if the stone is available. While it takes time and a lot of sweat, it represents a commitment to non–energy-consuming permanence.

Construction of a dry-laid wall begins with a layout of the wall, complete with staked perimeter. In ground-frost areas it may be necessary to excavate below the frost line, and this can, in the worst areas, double the final construction height. As a general rule, these walls should be half as wide at the base as they are high, and they should taper inward 3 inches for every vertical foot of rise. The stone-laying technique employed is two-over-one and one-over-two, the same as for any stone wall. Universal as this rule is, the building techniques vary with the size and shape of the stones being used.

Anybody can make a fence if all he has to work with are long, rectangular, flat-faced stones, but in the *real* world, all that seem to be available are rounded ones. They defy coursing (horizontal layering), and they do not lend themselves to neat "through stoning," which is placing a stone so that it shows on both the inner and outer faces of the fence. The fact of the matter is that walls made with

rounded stones are usually utilitarian affairs made by a farmer as he clears his fields. He usually has only three kinds of rounded stones to work with — little ones, big ones, and bigger ones — and, to complicate matters, he is usually working under seasonal pressure to get his seedbed ready. Despite these limitations and pressures, his handiwork has, with occasional spring repair, withstood the tests of time, weather, and countless forays by livestock.

Observing the "two-over-one, one-over-two" rule, the farmer tries to put the bigger stones at the base of his wall facing out, and he fills the space between with the little ones. As the wall rises, he uses the big ones as facing stones, while he continues to fill in with the little ones. When the fence reaches its predetermined height, the farmer caps it with the largest, flattest stones he can find. Stone wall fetishists (usually of the flatland variety) regard this kind of stone fence building with amused tolerance, but this only because they have regular, flat stones to work with, and nothing better to do with their time than to carp.

The assembled-masonry stone fence builders are divided into similar camps: those who contend that

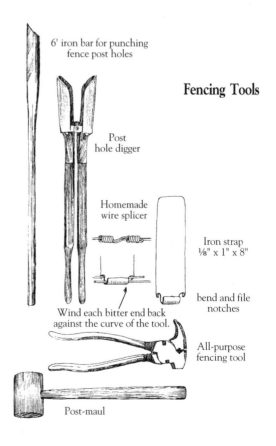

Fencing Tools

6' iron bar for punching fence post holes

Post hole digger

Homemade wire splicer

Iron strap ⅛" x 1" x 8"

bend and file notches

Wind each bitter end back against the curve of the tool.

All-purpose fencing tool

Post-maul

Stone Walls

Dry-laid walls need yearly upkeep.

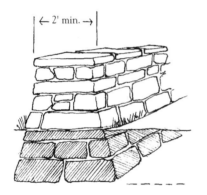

Hand-laid, mortared walls need little maintenance if tapered and footed below the frost line, but they need many specialized stones.

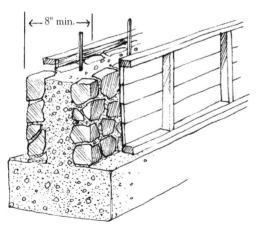

Slipformed stone and concrete walls have the same advantages as hand-laid mortared walls, but they require fewer materials, fewer specialized stones, less construction time, and only basic skills.

the only good masonry fence is one that is hand-laid, and those who make their fences with stone and concrete that is laid between forms. Laying stone with mortar or concrete, with or without benefit of forms, is saving of stone, and allows the builder to restrict the width of the wall to 8 to 24 inches. As a rule, hand-laid fences using cement must be wider than the formed wall, because the latter usually uses reinforced concrete. One-faced, formed, reinforced concrete fences having a footed wall can be as high as 8 or 9 feet, yet only be 8 inches wide. Two-faced walls should be at least 12 inches wide. The major drawback to building a formed, reinforced concrete fence is the high initial expense for materials — five or six times that of wire-and-metal-post fencing.

These cost factors often persuade small-scale farmers to erect wood post and galvanized wire fencing. Because of the rotting factor, wood posts are given the most attention. Most durable are juniper, split pitch pine, black locust, Osage orange, and gambel oak. Next in preference are cedar, red cedar, redwood, mulberry, catalpa, mesquite, desert-willow, honey locust, white oak, and sassafras. Last come those of limited durability, a durability that can be extended as much as three times with the application of a coal-tar creosote or other preservative: Douglas fir, ash, ponderosa pine, aspen, ash, beech, cottonwood, lodgepole pine, willow, box elder, hickory, maple, red oak, spruce, and tamarisk.

Building a solid farm fence using wire and wood is a subject worthy of a book in itself, but the general basic principles are:

1. Use plenty of fence posts. A spacing of 10 feet is a good average.
2. All corner and gate posts should be braced and secure. Where possible dig or cement them 3 or 4 feet into the ground.
3. Be sure to use the correct type and gauge of wire for your purpose.
4. Neatly stretch the wire to a tautness that will resist probing pressures by livestock or pests, but not so taut as to prohibit contraction in cold weather.
5. Set all corner and line posts before applying the wire or boards. Affix the wire or boards on the outsides of the posts *only* for appearance's sake, or when the sole intent is to keep pests out.

Typical Farm Gates

Typical Gate Closures

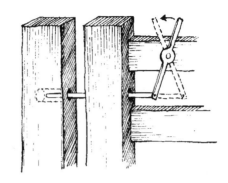

This simple closure is adequate for most livestock.

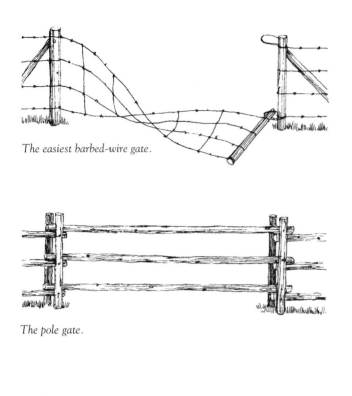

The easiest barbed-wire gate.

The pole gate.

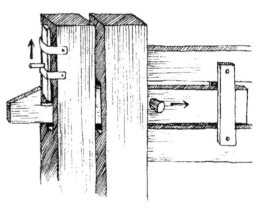

This closure thwarts clever goats.

The wheel-mounted stock gate works on level land only.

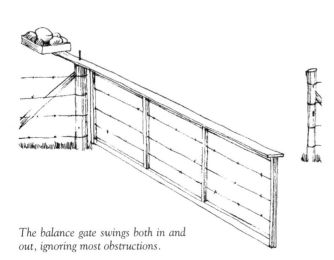

The balance gate swings both in and out, ignoring most obstructions.

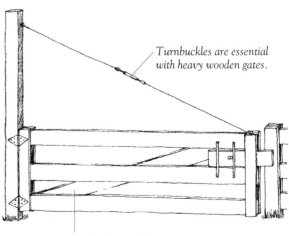

Turnbuckles are essential with heavy wooden gates.

Note the diagonal brace positioning.

THE FARM SHOP

Often the busiest, most-frequented spot on the small-scale farm is the shop. Here the farmer repairs old tools, creates new ones, and, in general, conducts his day-to-day business. A shelter full of tools, nuts and bolts, wood shavings, and the flotsam of every farm activity, the shop represents a vital and necessary part of every farm. As the nerve center of the farm, the shop is too important to be planned casually. This is especially true for small-scale farmers, who subscribe to a higher-than-usual degree of self-sufficiency, and to working with old, "outmoded" equipment that is in line with the scale of their ventures.

Shops, like farmers, differ greatly. The worst are jumbles of heaped rust surrounding a sawhorse, while others are studies in organization. Every farmer envies ones that sport labeled bins, spacious shelving, and orderly displays of tools above a tidy workbench, but more often than not his own workshop is a long way from this ideal.

To avoid a makeshift, messy shop where one can never find anything the farmer must analyze the functions of the shop, and plan it accordingly. Most farm shops divide into metal- and woodworking sections, and it pays to recognize this division. Each should have its own workbench and vise, and each should have storage spaces to hold the different kinds of raw materials.

The easiest way to plan a workshop is to build a structure to house it, but few farmers have the capital and time to invest in the project. In practice most have to make do with a corner of an existing barn, or, at best, with adding to an existing structure. The primary consideration in this planning is to make the shop usable year-round. In winter it should be heated. The heating source may be nothing more than a pot-bellied stove, but it is important, because winter is the only time when most farmers find time to "tinker" in the shop.

The next consideration is space. If the space one can give to a shop is limited, it should be given to the workbench area, but whenever possible this enclosed space should be expanded enough to accommodate a large piece of farm machinery. There is no substitute for working on a piece of machinery outdoors in foul weather to convince a farm shop planner to expand his work space. Tools are the last,

but not the least important consideration in planning the farm shop. Tools a farmer will need vary with his crops and his ability to "make do," but the following are the basic essentials:

Metal Shop

Hand tools:

wrenches — socket (set ⅛"–¾")	hand drill (Belly type)
wrenches — open, box (set ⅛"–1¾")	hammers (large and small ball peen)
wrenches — ignition (set)	sledgehammer (8 lb.)
wrenches — adjustable (various)	metal files (various)
wrenches — pipe	carborundum stones (various)
pliers — various	tinsnips
screwdrivers — various	pop-riveter
visegrip pliers	prybar
diagonals	anvil (with assorted anvil tools)
hacksaw	electric drill (½", with assorted bits)

Bench tools:

vise — machinist-type	bench grinder (hand or electric)

Other:

welder (arc or acetylene)	forge (small farrier-type)

Wood Shop

Hand tools:

hammers — claw type	wood files (various)
handsaws (rip and crosscut)	coping saw
framing square	pinch bar
measuring tape (retractable)	marking gauge
chisels (¼"–1½")	plumb bob
carpenter's level	chalkline
jack plane	100' tape on reel
putty knife	electric handsaw
nail set	electric saber saw
dividers (8")	vise — wood-type
brace (with bit set)	electric bench saw (arbor-type)

One can get along without the power tools (drills, grinders, and saws), but the shop work gets done much faster with them. Special jobs occurring over a span of several years will inevitably result in the addition of special tools. Pullers, torque wrenches, ring compressors, brake tools, tire repair tools, sickle-bar-tooth sharpeners, bolt cutters, bandsaws, drill presses, etc. will all eventually be acquired, and it is wise to look ahead and provide space for them in the planning stages.

Since most of today's farm equipment is constructed of metal, it is that part of the shop that sees the most use. In the past, one could get by with a forge and anvil for shaping and mending metal, but this is no longer the case. Much of the welding today has to be done "in place," that is, on the equipment, and this necessitates the acquisition of an arc or acetylene welding outfit and sometimes both. The welding jobs on thicker metal (½" and over) should be reserved for the arc welder, the smaller metal jobs for the acetylene rig. One of the big advantages the acetylene outfit has over the arc welder is that it is more easily transportable, a real plus when a piece of equipment breaks down beyond reach of an extension cord.

While the forge and anvil no longer enjoy the usage they once did, they are not merely subjects for an antique collector. There are still no better tools to use for bending, shaping, or tempering iron than these, especially when the process involves repeated heatings. Other than small amounts of coal or charcoal, these tools use only the farmer's energy, and they offer the farmer the unique capacity to create his own tools and repair parts from discarded iron.

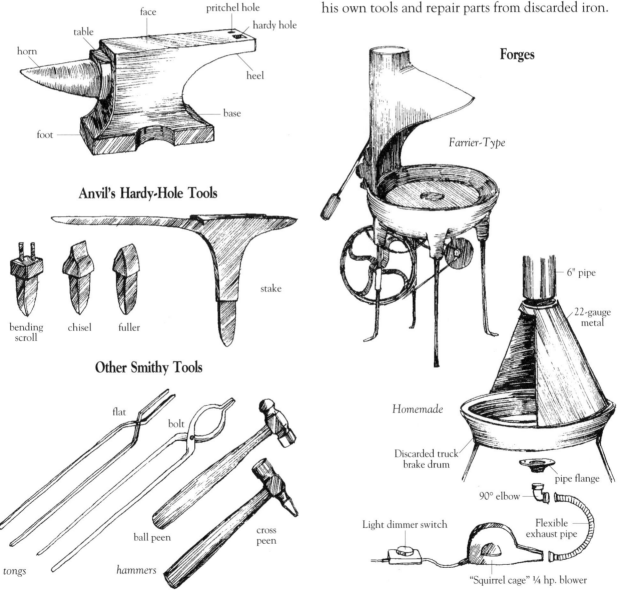

Anvil Parts

face
pritchel hole
hardy hole
table
horn
heel
base
foot

Anvil's Hardy-Hole Tools

stake
bending scroll
chisel
fuller

Other Smithy Tools

flat
bolt
ball peen
cross peen
tongs
hammers

Forges

Farrier-Type

6" pipe
22-gauge metal
Homemade
Discarded truck brake drum
pipe flange
90° elbow
Flexible exhaust pipe
Light dimmer switch
"Squirrel cage" ¼ hp. blower

CHAPTER 8

Conclusions

Farming is a state of mind, not an adopted avocation. If your life's objective is to make enough money from farming to become a part of the consumer culture — to amass *things* like new cars, automatic dishwashers, and color TVs — then small-scale farming is not for you. Given these priorities you are better off staying with a nine-to-five city job, or floating an enormous farm loan so that you can compete in the same arena with agribusiness. Simplicity of lifestyle, a dedication to careful planning, and a steadfast commitment to permanence are the essential elements of a successful small-scale farm venture.

Too many "futurists" (academic professional fortune-tellers) are wasting words on the polemics of agricultural change, and not enough are addressing the nuts and bolts of accomplishing the change. This book, while it lacks the exhaustive detail that is ultimately necessary, is intended to touch the basics of getting started with one's own small-scale farm. It will have served its purpose if only a few of its readers are encouraged to try their hand at farming, and are guided by the principles outlined in these pages.

At this writing, farming in the United States is in the midst of cataclysmic change, change that must ultimately resolve itself to the benefit of the small-scale farmer. Higher-priced energy, inflation, devaluation of the American dollar abroad, and the Frankensteinian character of corporate agriculture are the contemporary (and ongoing) economic realities that will force the price of agribusiness-produced food ever upwards, despite lackluster political efforts to depress them artificially. It is a gloomy fact that governmental efforts to control food prices end up costing the consumer through his tax bill, and lack of action ends up costing him at the checkout counter of the supermarket.

But there is no greater stimulus to the increased sales of small-scale–farm-produced food than that tasteless agribusiness product which currently dominates supermarket shelves. Using outmoded, "obsolete" equipment and archaic, labor-intensive practices, the small-scale organic farmer is bound to find his way into the consumer's pocketbook through his taste buds.

Agribusiness has proven to be a high-priced scenario. The 1980s saw big, debt-loaded farmers go under in droves. But, bad as it has been, the toll exacted of the farmer does not speak to the real costs of this evolutionary change. The land, exhausted by an unremitting pressure to produce high yields, and poisoned by the decades-old use of insecticides, herbicides, and fungicides, will take years to recover. Soil rebuilding efforts are, by themselves, awesome to contemplate. Just to replace a necessary part of the billions of tons of topsoil lost over the years through open-land agricultural practices will require sustained effort for a length of time at least twice that consumed in creating the problem.

There is no time like the present to begin, and no more likely place to start than with the establishment of the *non*-establishment small-scale farm. The newcomer to small-scale farming does not need to bring with him an anthropomorphic reverence for soil (though it probably makes as much sense as many religions), but he does need to find and nurture within himself a respect for the soil as a basis of his livelihood. He will soon find that it is a reciprocal arrangement — if he cares for his soil, it will, in turn, take care of him.

He also needs to bring with him an awareness of the benefits of smallness of scale. E.F. Schumacher, in his fine book, *Small is Beautiful*, says it well:

It is . . . obvious that men organized in small units will take better care of their bit of land or other natural resources than anonymous companies or megalomanic governments which pretend to themselves that the whole universe is their legitimate quarry.

But, as with any philosophical "ism," there is danger of overreaction:

Today, [reasoned Schumacher] we suffer from an almost universal idolatry of giantism. It is therefore necessary to insist on the virtues of smallness — where it applies. (If there were a prevailing idolatry of smallness, irrespective of subject or purpose, one would have to try and exercise influence in the opposite direction.)

A balance between scale and appropriate technology needs to be arrived at, and this will become the challenge of the future to the small-scale, polycultural farmer. Even today, acquiring and maintaining a sense of perspective is one of the most difficult tasks the small-scale farmer confronts. This task is made more difficult by the day-to-day exigencies of making ends meet. It would, for example, be fatuous and arrogant to insist (on platitudinous grounds) that newcomers to farming never use commercial fertilizers, or to condemn them for swapping their scythe and bull rake for a side-delivery rake and a baler.

Small-scale agriculture is already an established fact in many European countries, and we would do well to emulate some of their practices. England and France, for example, are making significant efforts, with government support, to establish biological controls to replace (at least in part) insecticides. They are also encouraging the design and manufacture of equipment that is calculated to work small acreages more efficiently. This forward-looking policy is matched by the concerted effort to protect farmlands by law from the threat of development. They are getting results, and among the more startling is the demographic study which shows an increase in the amount of land devoted to growing food, and an increase in the number of farmers working the land.

This country has, unfortunately, not been as kind or as foresighted. Here, small-scale farmers are thrown back onto their own resources. If, as is often the case, these resources are slim, most young land buyers surrender the idea of buying an ongoing farm, and settle for raw acreage. This represents a long-term investment, for all the structures need to be built from scratch. There are, however, advantages to this plan: One can develop the land and structures as he acquires capital, and the construction costs can be ameliorated by the buyer who uses his own labor. All of this requires time.

Those who do not have enough capital to afford to purchase raw acreage outright face an even longer period of "catching up." In this situation, the would-be farmer may be forced to assume a mortgage and figure the interest he pays the banker into his overhead. This is unproductive, and should be adopted only as a last resort. Fortunately, there are other alternatives available.

One option, for example, is to approach a prospective seller with a swap or barter proposition. Old-timers are particularly amenable to these kinds of deals, because they have had first-hand experience with them, and they value skills and services in a way that is rarely known today. Exercising this kind of option usually requires that the land seeker become a part of the farming community beforehand.

A second option is acquiring land in common with other land seekers. The reasoning behind this is that larger acreages generally are sold for a smaller price per acre. There are two ways to approach this option: the first is to buy outright, along with people of your own choosing, and the second is to lease (long-term) from a fund-pooling land trust, of which you are a voting member.

Because it is easier to organize, and the result is traditional outright ownership, the first method enjoys the most usage. It is important with all common land-buying arrangements of this sort to contractualize the ownership between partners, even though the parties involved may be close friends or even relatives. This parceling out should be done after a professional survey.

The second method, forming or joining a cooperatively owned and/or cooperatively run land trust, is more complicated, and it generally involves more social commitment. Land trusts can vary in philosophy from profit-making ventures to idealistic communes, but the type that should interest today's would-be farmer are those established as nonprofit organizations with stewardship overtones.

Several such trusts have been successfully established across the country. The land seeker may be interested in pursuing this option after studying the sample articles of incorporation and sample lease that appear on pages 125–127 in the Appendix. These samples are used by the Earth Bridge Community Land Trust of Putney, Vermont, and represent only one of many approaches to this concept.

The Earth Bridge Trust is incorporated and members call their land trust a "project in cooperative self-reliance." They buy large parcels of land at lower prices (in some cases the land is donated) and remove it from the speculative land market. To lessees they offer eighty-nine–year leases that are inheritable and renewable, but nonspeculative. That is, lessees own whatever buildings or improvements they put on the land, and may sell them, but they may not add the value, speculative or not, of the land to the sale price. Using his lease payments, the lessee pays off the land trust's original purchase price of the land, and then the lease payments cease. He continues to pay the property taxes on the real estate, and he is, for all intents and purposes, the owner-by-use of the land. As a member of the land trust, he has voting rights in deciding future trust policy.

Husbandry-oriented organizations like Earth Bridge are a welcome way for new, would-be farmers to get on the land. Lacking the necessary capital to buy outright, the farmer gets on the land with payments that are like rent, and with the cheerful prospect that the "rent" ceases when the value of the land is paid.

It is fascinating that the method the farmer uses to get on the land, whether it be rent, inheritance, marriage, or sweat, does not affect his ultimate relationship with the soil. This relationship defies rationalization and scientific scrutiny. What other continuing line of work calls for such large financial investment with such risky and minimal returns? What employer would have the gall to offer such prolonged hours of sweaty and often dangerous work, and ignore the weather exposure, lack of medical coverage, lack of retirement provisions, and absence of job security? Using these criteria, farming makes no sense at all. Yet a farmer will, if given the slightest encouragement, carry on his love affair with the soil until the soil finally claims him in the grave.

Farmers are not a glib lot. When asked why he continues when a worse year follows a bad one, any given farmer is apt to respond with a wry "win some, lose some," or launch into a delineation of his family's roots on his farm. But the same farmer will also tell you that it only takes one growing season to put down roots. This is borne out by the fact that newcomers to farming are as bewitched by this link to the soil as the fourth generation of old-timers.

Perhaps the answer lies in the few seconds the farmer finds at the end of a punishing day, when he surveys what he has done and finds it good. He knows then that there is no more satisfying thing to do in the world than to put seed in the ground, and then to sit back and watch it grow.

Handy Hints, Formulas & Trivia

THE ACRE

ACRE: "a measure of land containing 160 square rods or perches, or 4,840 square yards." *Webster's, 2nd Ed.*
— the original meaning was an open, plowed, or sowed field.
— from the Greek *agros*, meaning 'field'.

Table showing one side of a square tract of land containing:

$\frac{1}{10}$ acre	= 66 feet per side	= 4,356 square feet
$\frac{1}{8}$ acre	= 73.8 feet per side	= 5,445 square feet
$\frac{1}{6}$ acre	= 85.2 feet per side	= 7,260 square feet
$\frac{1}{4}$ acre	= 104.4 feet per side	= 10,890 square feet
$\frac{1}{3}$ acre	= 120.5 feet per side	= 14,520 square feet
$\frac{1}{2}$ acre	= 147.6 feet per side	= 21,780 square feet
$\frac{3}{4}$ acre	= 180.8 feet per side	= 32,670 square feet
1 acre	= 208.7 feet per side	= 43,560 square feet
1½ acres	= 255.6 feet per side	= 65,340 square feet
2 acres	= 295.2 feet per side	= 87,120 square feet
2½ acres	= 330 feet per side	= 108,900 square feet
3 acres	= 361.5 feet per side	= 130,680 square feet
5 acres	= 466.7 feet per side	= 217,800 square feet
10 acres	= 660 feet per side	= 435,600 square feet

One Acre Equals
— 0.4047 hectare
— 43,560 square feet
— 160 square rods
— 4,840 square yards
— 10 square chains
— 4 roods
— $\frac{1}{640}$th of a square mile

SURVEYOR'S MEASURE
7.92 inches — 1 link
25 links — 1 rod
4 rods — 1 chain
10 sq. chains or 160 rods — 1 acre
640 acres — 1 sq. mile or 1 section
36 sq. miles (6 miles sq.) — 1 township

Homemade Surveying Tools

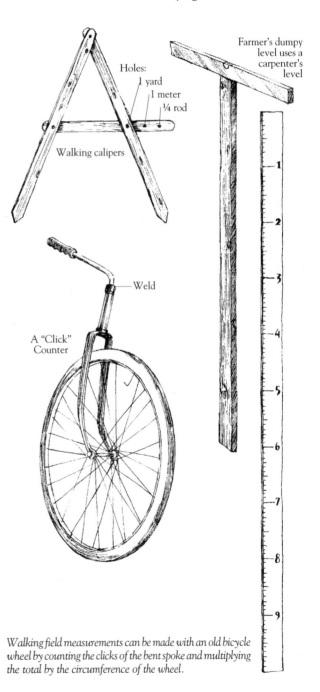

Farmer's dumpy level uses a carpenter's level

Holes:
1 yard
1 meter
¼ rod

Walking calipers

Weld

A "Click" Counter

Walking field measurements can be made with an old bicycle wheel by counting the clicks of the bent spoke and multiplying the total by the circumference of the wheel.

COMMON MEASURES

Long Measure

12 inches .... 1 foot
3 feet .... 1 yard
5½ yards .... 1 rod
320 rods............................. 1 mile
1 mile .... 5,280 feet
1 size (shoe) .... ⅓ inch
1 hand .... 4 inches
1 fathom .... 6 feet
1 knot .... 1.15 miles

Dry Measure

2 pints .... 1 quart
8 quarts .... 1 peck
4 pecks .... 1 bushel

1 bushel contains 2,150.42 cubic inches or approximately 1¼ cubic feet.

Staking a contour line using a carpenter's level.

Land Measure — Length

7.92 inches = 1 link
25 links = 1 rod • 16.5 feet • 5.5 yards (1 rod = 1 pole = 1 perch)
4 rods = 1 chain (Gunter's) • 66 feet • 22 yards • 100 links
10 chains = 1 furlong • 660 feet • 220 yards • 1,000 links • 40 rods
8 furlongs = 1 mile • 5,280 feet • 1,760 yards • 8,000 links • 320 rods • 80 chains

Land Measure — Area

30¼ square yards = 1 square rod • 272¼ square feet
16 square rods = 1 square chain • 484 square yards • 4,356 square feet
2½ square chains = 1 rood • 40 square rods • 1,210 square yards
4 roods = 1 acre • 10 square chains • 160 square rods
640 acres = 1 square mile • 2,560 roods • 102,400 square rods
1 section land = 1 square mile • 1 quarter section • 160 acres

Ropes and Cables

2 yards = 1 fathom

120 fathoms = 1 cable's length

Liquid Measure

4 gills = 1 pint = 16 fluid ounces
2 pints = 1 quart = 32 fluid ounces
4 quarts = 1 gallon = 32 gills = 8 pints
1 gallon contains 231 cubic inches.
1 cubic foot contains 7½ gallons or 7.4805 gallons (liquid measure)

Miscellaneous

3 inches = 1 palm   18 inches = 1 cubit
4 inches = 1 hand   21.8 inches = 1 Bible cubit
9 inches = 1 span   2½ feet = 1 military pace

AREAS OF PLANE FIGURES

Area of a square = side x side

Area of a rectangle = length x height

Area of a triangle = ½ the base x the height

HAND FILES

The names of the principal files are: hand, mill, flat, pillar, warding, square, round, half-round, three-square, and knife. They are graded by the coarseness or fineness of their cut, and come in the following gradations:

No. 000 cut	—	coarse cut
No. 00 cut	—	bastard cut
No. 0 and No. 1 cuts	—	second cut
No. 2 and No. 3 cuts	—	smooth cut
No. 4 and No. 5 cuts	—	dead-smooth cut

BOARD FEET IN A LOG

Subtract 4 inches from the diameter and square the remainder. The result will be the number of board feet in a 16-foot log. Add ⅛ for 18-foot logs, ¼ for 20-foot logs. Subtract ⅛ for 14-foot logs, and ¼ for 12-foot logs.

CORD WOOD

A standard cord is a solid measure equivalent to 128 cubic feet. In cordwood operations, the agreement is that a pile that is 8 feet long, 4 feet wide, and 4 feet high is a cord. Thus, to find the number of cords in a pile of wood: Multiply the length, width, and height together — the product will be cubic feet, which are then reduced to cords by dividing by 128.

The term "face cord" is an unreliable measure because it has too many definitions. Generally it is meant to denote any stack of wood measuring 4 feet by 8 feet on its face — the depth of the stack (individual piece length) can be 12 inches (¼ cord), 16 inches (⅓ cord), or whatever.

The most disputed cord measurement is the one called the "throwed-on" cord, or the "running" cord. Because stacking wood is time-consuming (thereby cutting into the profits), many firewood dealers use a cubic foot measure to sell wood that is applied to a randomly "throwed-on" pile of wood; 160 cubic feet is a typical "throwed-on" cord.

ELECTRICITY

A watt is a unit of electrical power that is determined by multiplying volts (pressure) by amperes (rate of flow).

An ohm is a unit of electrical resistance. It allows one ampere to flow at one volt of pressure.

Energy note: Four 25-watt bulbs use the same amount of current as one 100-watt bulb, but the latter gives 50 percent more light.

HEAT MEASUREMENT

A calorie is the amount of heat required to raise the temperature of one gram of water from 14.5°C. to 15.5°C.

A Btu (British thermal unit) is the quantity of heat needed to heat a pound of water from 39°F. to 40°F.

One Btu equals 252 calories.

MECHANICAL HORSEPOWER

A mechanical horsepower is equal to 33,000 pounds raised one foot in one minute, or its equivalent.

One horsepower is equal to 746 watts.

Pulling Posts

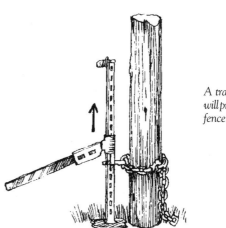

A tractor jack and chain will pull the most stubborn fence posts.

MANURE

Generally speaking, 1 ton of cow manure equals 200 pounds of 10–5–10 commercial fertilizer.

PULLEY SIZES

PROBLEM: What is the right **driven** pulley diameter to use if you want 3,000 rpm when you know, 1) that the power source rpm is 2,000, and 2) that it is mounted with a driving pulley having a diameter of 12 inches?

ANSWER: Multiply the diameter of the driving pulley by its rpm, and then divide the answer by the desired rpm.

EXAMPLE:

12 x 2,000 = 24,000; 24,000 ÷ 3,000 = 8-inch pulley

This may be worked backwards to arrive at a desired driving pulley size.

EXAMPLE:

8 x 3,000 = 24,000; 24,000 ÷ 2,000 = 12-inch pulley

BALER TWINE

Baler twine is usually made of sisal fiber, and is stronger than old-fashioned binder twine. Brands in use today commonly have a tensile strength of 325 pounds, and contain about 230 feet of twine length per pound. By contrast, binder twine had a strength of 90 pounds and contained 500 feet per pound. On a typical baler, a 43-pound bale (2 rolls) of baler twine having 10,000 feet can tie 550 bales that are 30 to 36 inches long.

GLOBES AND CIRCLES

To find the length of the diameter of a circle, multiply the circumference by 0. 31831.

To find the length of the circumference of a circle, multiply the diameter by 3.1416.

To find the area of a circle, multiply the square of the diameter by 0.7854.

To find the surface of a ball, multiply the square of the diameter by 3.1416.

To find the number of cubic inches in a ball, multiply the cube of the diameter in inches by 0.5236.

WOOD FUEL

The average farm woodlot produces about 1 lean cord of firewood per acre per year. It is better to figure on ¾ cord per year. The following hardwoods are listed according to their efficiency and desirability as fuel wood.

BOARD MEASUREMENT

The unit of measure is the board foot, which is a board 1 inch thick and 1 foot square. Lumber is usually sold in units of 1,000 feet, board measure (B.M.).

To find B.M., multiply the length in feet by the width and thickness in inches, and then divide the product by 12.

FARM WOODS

Farmers need to know what woods are best suited for specific jobs. The most common need is for wood that will endure wet conditions. Fence-post woods like cedar, locust, cottonwood, juniper, cypress, or hop hornbeam are also usable for lumber or for barn posts, but tamarack (larch), hemlock, or lodgepole pine are more commonly used. Pine, fir, and spruce are generally employed for framing or cabinetry, and the hardwoods like maple, birch, oak, black gum, beech, and ash are used for flooring. Hickory and white ash are used for helves, snaths, tool handles, and eveners. Poplar (aspen) or basswood make serviceable wagon boxes (also birdhouses), and they are the woods most sought after for use around foods — because they impart little or no taste to the food. Oak is the workhorse of woods used around the farm. Heavy machinery and equipment that takes hard wear call for oak's toughness, durability, and strength.

HEAT EQUIVALENTS OF SEVERAL COMMON FIREWOODS

Species	Btus in 1 cord	No. 2 fuel oil (gallons)	Anthracite coal (tons)	Natural gas (100 cu. ft.)
Hickory (shagbark)	24,600,000	251	1.12	308
Apple	23,877,000	244	1.09	298
Oak, white	22,700,000	232	1.04	284
Beech	21,800,000	222	.99	273
Oak, red	21,300,000	217	.97	286
Maple, sugar	21,300,000	217	.97	286
Birch, yellow	21,300,000	217	.97	286
Ash, white	20,000,000	204	.91	250
Birch, white	18,900,000	193	.86	236
Maple, red	18,600,000	190	.84	232
Elm, American	17,200,000	176	.78	215
Aspen (popple)	12,500,000	128	.57	156
Pine, white	12,022,000	123	.55	150

CEMENT MIXER

Besides mixing cement for the innumerable construction jobs around the farm, the cement mixer is useful to mix hayland seed blends, seed and inoculant, and chicken scratch.

Stoneboat

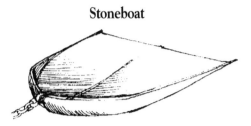

An old, discarded auto hood makes a fine stoneboat.

LOOSENING RUSTY BOLTS

A good penetrating oil is indispensable for working around old farm machinery. Commercial preparations are handy, but expensive. Try mixing 1 tablespoon butyl alcohol into 2 tablespoons kerosene, then stirring this mixture into 7 tablespoons mineral oil. This homemade concoction should serve just as well as commercial preparations.

In a pinch, Coca-Cola will dissolve rust, but it will not oil the threads.

If all else fails, use a tool called a "nutcracker," or hacksaw down through the nut. Then squeeze the halves together, drop a little oil in, and back the nut off.

RUST REMOVER

Mix the Following:

    1 tablespoon ammonium citrate crystals
    2 cups water

WOUND COATING FOR TREES AND SHRUBS

Mix the Following:

    1 cup zinc oxide
    2 cups mineral oil

CATERPILLAR AND INSECT TREE BANDS

Mix the Following:

    1½ cups powdered rosin
    1 cup linseed oil
    1 tablespoon melted beeswax

FIGURING TREE HEIGHTS

Set up a stick and measure its shadow. Measure the length of the tree's shadow. The length of tree's shadow, times the height of the stick, divided by the length of the stick's shadow equals the height of the tree.

ACREAGE COVERED BY FARM MACHINES

1. Divide the width of cut in inches by 100 to determine the acres covered in 1 mile of travel.
2. Multiply the width of cut in inches by miles per hour of tractor speed, and then multiply the result by tractor working hours. Finally, divide this result by 100 to get the number of acres covered in an hour.
3. Multiply the width of cut in feet by miles per hour of tractor speed to determine the number of acres covered in a ten-hour day.

BUYING HAY

Hay is normally sold in one of three categories: "behind the baler," "in the barn," or delivered. The price the buyer can be expected to pay rises progressively with each of these categories. Depending upon the quantity to be bought, the hay may be sold by the bale or by the ton. Individual bales may vary in weight, depending upon length of bale, when cut, kind of hay, or degree of dryness. Here are some hints for getting good hay value:

1. When possible, buy "behind the baler" hay so that you can see how the hay was processed to that point. Remember that freshly baled hay will weigh more than hay that has had time to dry in the barn.
2. To avoid needless hassle, agree beforehand whether the price for the hay is by the bale or the ton, and determine (if no scales are available) how many bales constitute a ton. It is not rude to bring spring scales to weigh a few bales to get an average weight.
3. If buying hay "in the barn," or delivered, look, smell, and drop it before you buy, or be very sure of the seller's reputation.

  • Look for green color, about the color of your hard-earned dollar bill. Dark hay is late cut or moldy.
  • Look out for large-stemmed, rank hay. The goodness in hay is in its leaf, not its stem.
  • Smell the hay, and know what mold or mildew smells like. If you have hay fever or a cold bring a friend.

HAY PRICE AND WEIGHT TABLE

Weight of Bale (lbs.)	No. of Bales per Ton	$ Price per Ton if Price per Bale Is						
		.50	.75	1.00	1.25	1.50	1.75	2.00
40	50	25.00	37.50	50.00	62.50	75.00	87.50	100.00
50	40	20.00	20.00	40.00	50.00	60.00	70.00	80.00
60	33⅓	16.67	25.00	33.33	41.66	50.00	58.33	66.66
70	28½	14.25	21.38	28.50	35.63	42.75	49.88	57.00
80	25	12.50	18.75	25.00	31.25	37.50	43.75	50.00

- Drop one or two bales. They should bounce slightly if dropped on a corner of the bale. A bale that is moldy inside will thud warningly.

FARM SOCIAL AFFAIRS

Dancing

SQUARE DANCES: Invitations should guarantee enough for at least one square (four couples), and a caller. Music is traditionally provided by a fiddler and whatever accompaniment happens to come along. The key to a successful square dance is the caller, though some think it is the liquid refreshment. Square dances can be combined with any of the victual occasions cited below.

ROUND DANCES: This is a kind of dancing commonly referred to by city folks as ballroom dancing. No caller is needed, and the music for the affair can vary from that provided by one instrument to that of a whole orchestra.

Victual Occasions

BOX SOCIAL: This is a fund-raising/social affair that is traditionally put on by the local Grange, PTA, or women's fellowship. It is an auction with male chauvinist overtones. As with most victual events, the work is traditionally done by women. The rules go like this:

- Women cook a meal big enough for two or more persons, and put the meal and their name into a decorated box.
- The box is then auctioned off to the highest male bidder, and the buyer gets the meal and the company of the cook during the consumption of the meal.
- Traditionally, the woman surreptitiously lets the "right person" know the wrapping details of her box, but the competitive auction

atmosphere often pits the bidding of a confirmed bachelor who is interested in the food against that of a lovesick swain who is after the company of the cook.

CHICKEN PIE, TURKEY, AND POT-LUCK SUPPERS: Many victual happenings are tied to fund-raising, and these occasions are no exceptions. Paid admission, with the proceeds going to a worthy cause, is the order of the day. All such suppers, except the pot-luck, are prepared and served by the organization sponsoring the affair.

Parties

KITCHEN JUNKETS OR SOIRÉES: These are parties given in private homes where the giver usually extends invitations. The occasion may be a special occasion, like an anniversary or a birthday, or it may simply be given "because it's been a long winter." Refreshments and occasionally a meal are served, but the primary purpose is to dance — either square or round. For the pleasure of the dancers, it is traditional to sprinkle cornmeal or sugar on the kitchen floor.

SHIVAREES: A shivaree is a party reluctantly given by host and hostess. It is occasioned by a wedding — specifically it occurs the night of the wedding. Celebrants gather in the middle of the night at the home of the newlyweds. They are armed with noisemakers like cowbells, pots, and pans, and they raise a bedlam until they are finally invited in, fed, and refreshed. Rambunctious male celebrants have on such occasions been known to stuff the marital couch with whole oats, or even to steal the bride. Oafish, but all in fun.

HAYRIDES: Typically given in the summer or late fall, the hayride is an occasion for quiet spooning by young, or not so young, adults. It usually ends with hot refreshments at the home of the host.

Singing

SACRED HARP SINGING: This is an old-time New England and Appalachian custom that has recently been revived. It entails a gathering of folks (in home or hall) to sing. In the old days they mostly sang sacred hymns. This is a participatory affair, not a spectator one. The singers sit in a square with basses opposite sopranos, and tenors opposite altos. Hymn or song numbers are suggested from a common songbook, and each song is sung through twice. The first singing is by syllable (do-re-me, etc.) in harmony. The second singing includes the words.

Working/Social

CORN-HUSKING PARTY: The host is the family whose corn is to be husked. Food and refreshments are served. The outstanding rule of particular interest to young people is that of granting kissing privileges to the person (boy or girl) who husks a red ear of corn. Hosts have been known to "salt" the cornbin with plenty of red ears that were grown specifically for that purpose. Little corn is husked today, thus this party is mostly a historical curiosity.

SEWING, QUILTING BEES: These are working parties for women. The product of their labors is used for various "good work" projects.

BARN-RAISING BEES: Fewer barns are constructed today than in the old days, but this idea is still employed in roof-raising on new houses. The men come to work on the structure of the host, and the women prepare a pot-luck meal. The host is responsible for the cider or any other work-inciting beverage.

LIGHTNING FACTS

- Four out of every five human deaths or injuries from lightning occur in rural areas.
- Lightning causes about $12 million worth of damage to farm structures each year.
- Wire fencing will carry lightning for 2 miles or more, unless grounded.
- If you hear thunder, the lightning is less than 20 miles away. If you see the flash, you can estimate its distance from you by counting the seconds until you hear the resultant thunder. Multiply the seconds by 1,100 to get the distance in feet (5 seconds equals approximately 1 mile).
- If caught outside during a lightning storm:
  - Stay away from lone trees, hilltops, fences, and ponds.
  - Do not ride in open equipment like a tractor.
  - Get out of fields or other open areas.
  - Hurry to a large structure or a closed truck or automobile.
  - If no vehicles or buildings are close, take cover in large clumps of trees or in a ditch or depression in the ground. Avoid being the outstanding bump on any topographic profile.

Charts, Tables & Addenda

CROPS

PLANT FOOD CONTENT OF VARIOUS CROPS AND AGRICULTURAL PRODUCTS

Crop or product	Yield or amount, air-dry	Nitrogen (N), lb	Phosphate (P_2O_3), lb	Potash (K_2O), lb	Lime (100% $CaCO_2$), lb
Corn, grain	60 bu	57	23	15	1
Fodder	2 tons	38	12	55	46
Total	—	95	35	70	47
Oats, grain	50 bu	35	15	10	4
Straw	1.25 tons	15	5	35	22
Total	—	50	20	45	26
Wheat, grain	30 bu	35	16	9	1
Straw	1.25 tons	15	4	21	14
Total	—	50	20	30	15
Barley, grain	40 bu	35	15	10	3
Straw	1 ton	15	5	30	16
Total	—	50	20	40	19
Rye, grain	20 bu	22	9	7	1
Straw	1.5 tons	15	8	26	21
Total	—	37	17	33	22
Kafir, grain	50 bu	43	16	10	2
Fodder	3 tons	62	17	88	56
Total	—	105	33	98	58
Sweet sorghum, fodder	4 tons	81	25	132	98
Tobacco,	1,500 lb				
Leaves	—	55	10	80	50
Stalks	—	25	10	35	35
Total	—	80	20	115	85
Cotton, lint	500 lb				
Seed	1,000 lb	38	18	14	3
Stalks, etc.	1,500 lb	27	7	36	80
Total	—	65	25	50	83
Soybeans, grain	25 bu	110	35	40	8
Straw	1.25 tons	15	5	20	60
Total	—	125	40	60	68
Clover seed	100 lb	2.9	1.9	1.5	1.0
Alfalfa seed	100 lb	5.87	1.16	1.17	0.45
Cowpea seed	100 lb	3.78	1.10	1.76	0.25

Continued

Crop or product	Yield or amount, air-dry	Nitrogen (N), lb	Phosphate (P_2O_3), lb	Potash (K_2O), lb	Lime (100% $CaCO_2$), lb
Forages (air-dry):					
Timothy	1 ton	26	10	30	13.5
Lespedeza	1 ton	43.5	10	23.5	49.5
Red clover	1 ton	40	10	35	60.5
Sweet clover	1 ton	37	9	33	73.5
Alfalfa	1 ton	46	12	45	71.5
Cowpeas	1 ton	62.5	12.5	45	56.5
Livestock and livestock products:					
Fat cattle	1,000 lb	25.0	16.1	2.4	32.0
Fat hogs	1,000 lb	18.0	6.6	13.0	11.3
Fat lambs	1,000 lb	20.0	11.2	1.7	23.5
Milk	1,000 lb	5.6	2.0	1.6	3.0
Butter	1,000 lb	1.6	—	—	—
Eggs	100 doz (162.5 lb)	3.5	1.7	0.8	26.2
Chickens	100 lb	3.5	0.85	—	2.8
Fruit and vegetables:					
Apples, fruit	400 bu	20	7	30	5
Asparagus spears	5,000 lb	20	6	38	—
Beans or peas:					
Seed	30 bu	73	23	24	9
Straw	—	22	7	31	—
Total	—	**95**	**30**	**55**	—
Beets:					
Roots	20,000 lb	36	14	34	—
Tops	—	40	16	62	—
Total	—	**76**	**30**	**96**	**55**
Blackberries, fruit............	4,000 lb	6	4	8	—
Carrots, whole crop	30,000 lb	120	50	240	—
Cabbage, whole crop	15 tons	100	25	100	30
Cantaloupe, fruit	4,000 melons	57	16	100	—
Cherries, fruit	8,000 lb	18	6	22	—
Grapes:					
Fruit	6,000 lb	8	4	15	—
Leaves and canes	—	13	4	13	—
Total	—	**21**	**8**	**28**	—
Lettuce, whole crop	15,000 lb	41	17	71	—
Onions, whole crop	600 bu	42	12	30	27
Peaches:					
Fruit	500 bu	30	15	55	7
Leaves and wood	—	55	10	45	—
Total	—	**85**	**25**	**100**	—
Pears, fruit	400 bu	18	6	17	—
Plums, fruit	8,000 lb	15	2	20	—

Continued on page 102

Crop or product	Yield or amount, air-dry	Nitrogen (N), lb	Phosphate (P$_2$O$_3$), lb	Potash (K$_2$O), lb	Lime (100% CaCO$_2$), lb
Potatoes (Irish):					
Tubers	300 bu	65	25	115	13
Tops	—	60	10	55	—
Total	—	**125**	**35**	**170**	—
Potatoes, sweet:					
Roots............................	300 bu	45	15	75	13
Vines............................	—	30	5	40	—
Total	—	**75**	**20**	**115**	—
Raspberries, fruit...................	3,000 lb	5	15	11	—
Spinach, tops........................	12,000 lb	60	20	30	31
Strawberries, fruit	180 crates	9	6	18	—
Tomatoes:					
Fruit	20,000 lb	60	20	80	25
Vines............................	—	40	15	95	—
Total	—	**100**	**35**	**175**	—
Turnips, roots	400 bu	51	31	69	29

PLANT FOOD REMOVED BY CROPS

Crops	Yields	Nitrogen (N)	Phosphoric Acid (P$_2$O$_5$)	Potash (K$_2$O)	Lime (CaO)
Alfalfa hay............................	5 tons	(238.)*	54.	223.	260.
Barley grain	40 bu @ 48	35.3	16.3	14.2	1.1
Barley straw	1 ton	11.2	3.6	24.	6.5
Barley, total crop	—	**46.5**	**19.9**	**38.2**	**7.6**
Beet-sugar roots	15 tons	78.	24.	96.	11.
Bluegrass (Ky.)	2 tons	53.2	21.6	84.	17.
Buckwheat grain...................	30 bu @ 50	21.8	15.	10.5	.4
Buckwheat straw	.75 ton	12.5	1.95	16.95	14.3
Buckwheat, total crop	—	**34.3**	**16.95**	**27.45**	**14.7**
Cabbage heads	15 tons	105.	21.	87.	51.
Clover hay (red)	2 tons	(82.)*	15.6	65.2	87.
Clover hay, alsike	2 tons	(82.)*	28.	69.6	55.
Clover hay, crimson	1 ton	(45.)*	12.	45.	55.
Clover, lespedeza	1 ton	(39.)*	21.0	11.0	57.
Corn grain	65 bu @ 56	59.	25.11	14.56	1.
Corn stover	1.75 tons	33.	15.75	45.15	17.

*Approximately 70% of the N in inoculated legumes is fixed from the air.
Source: Data from *Missouri Balanced Farming Handbook*, University of Missouri Agricultural Extension Service BF-5604. Analyses from various sources will vary considerably, but these probably best fit average Missouri conditions. Crops grown on better soils normally have higher composition than those grown on poorer soils.

Crops	Yields	Nitrogen (N)	Phosphoric Acid (P$_2$O$_5$)	Potash (K$_2$O)	Lime (CaO)
Corn cob	900 lbs	3.	.63	5.9	.1
Corn, total crop	**4 tons**	**95.**	**41.49**	**65.61**	**18.1**
Corn (for silage)	12 tons	81.6	38.4	105.6	19.
Cotton lint	500 lbs	1.5	.3	2.5	.8
Cottonseed	1,000 lbs	31.5	15.	15.	2.5
Cotton, total crop	—	**33.**	**15.3**	**17.5**	**3.3**
Cowpea hay	2 tons	(124.)*	38.4	165.2	50.
Flax grain	15 bu @ 60	31.7	13.5	8.5	2.8
Flax straw	.9 ton	20.6	3.4	18.9	13.
Flax, total crop	—	**51.3**	**16.9**	**27.4**	**15.8**
Hemp (dry stalks)	3 tons	20.	4.	44.	42.
Millet hay (common)	3 tons	80.	21.6	129.	23.
Oat grain	50 bu @ 32	31.7	13.	9.	4.5
Oat straw	1.25 tons	44.5	5.25	37.5	10.5
Oats, total crop	—	**46.2**	**18.25**	**46.5**	**12.**
Onions (bulbs only)	500 bu @ 56	60.	25.2	61.6	44.
Pea grain	20 bu @ 60	(44.)*	10.	12.1	2.8
Pea straw	1.5 tons	(30.)*	2.5	26.4	59.6
Peas, total crop	—	**(74.)***	**12.5**	**38.5**	**62.4**
Peas, green (total crop)	7.5 tons	(85.)*	18.	48.	76.5
Peanuts (shelled)	1,000 lbs	(48.8)*	10.2	6.5	—
Peanut shells	500 lbs	(6.)*	.7	4.0	—
Peanut hay	500 lbs	(15.2)*	2.5	16.	—
Peanuts, total crop	—	**(70.0)***	**13.4**	**26.5**	—
Potatoes, Irish	200 bu @ 60	42.0	14.4	63.6	3.4
Potatoes, sweet	200 bu @ 50	29.	9.	51.0	—
Rye, grain	25 bu @ 54	25.5	9.5	7.70	.7
Rye, straw	1.25 tons	12.0	7.00	19.75	7.7
Rye, total crop	—	**37.5**	**16.85**	**27.45**	**8.4**
Sorghum fodder, dry	4 tons	19.	8.6	32.8	—
Soybeans, grain	20 bu @ 60	(70.0)*	16.44	29.64	3.1
Soybeans, straw	1 ton	(35.0)*	2.4	17.8	45.0
Soybeans, total crop	—	**(105.)***	**18.84**	**47.44**	**48.1**
Timothy hay	2 tons	39.6	12.4	54.4	14.1
Tobacco leaves	1,500 lbs.	41.0	12.	96.0	58.0
Tobacco stalks	1,250 lbs.	26.0	18.75	75.00	9.6
Tobacco, total crop	—	**67.0**	**30.75**	**171.00**	**67.6**
Turnips, roots	15 tons	66.0	39.0	87.00	21.1
Wheat, grain	30 bu @ 60	35.6	15.48	9.54	1.0
Wheat, straw	1.6 tons	46.0	4.16	23.68	6.0
Wheat, total crop	—	**51.6**	**19.64**	**33.22**	**7.0**

The data are taken from averages of many analyses of common crops. The amounts are based on yields shown for productive soils. Lower yields will remove proportionally less. The data form a basis for calculating the fertilizer requirements for soils for each crop given. The nitrogen for legumes given is in parentheses, as much of it may come from the air. Amounts are given in pounds unless otherwise indicated.

SEEDS REQUIRED TO SOW 100 YARDS OF ROW[1]

Asparagus ....8 ounces	Corn ....1 pint	Onions, for sets ....6 ounces
Beans, bush.....................3 quarts	Cress ....4 ounces	Parsley...............................2 ounces
Beans, lima ....3 pints	Cucumber ....4 ounces	Peas....................................2 ounces
Beans, pole ....1 pint	Eggplant ....½ ounce	Pepper...............................½ ounce
Beet ....4 ounces	Endive ....2 ounces	Pumpkin ....2 ounces
Broccoli ....½ ounce	Leek ....2 ounces	Radish...............................6 ounces
Brussels sprouts...............½ ounce	Lettuce ....2 ounces	Rhubarb ....4 ounces
Cabbage ....1 ounce	Melon, water ....2 ounces	Salsify...............................4 ounces
Carrot ....3 ounces	Melon, citron ....2 ounces	Spinach ....6 ounces
Cauliflower ....½ ounce	Mustard ....4 ounces	Squash ....3 ounces
Celery ....3 ounces	Okra..............................12 ounces	Tomato ....1 ounce
Collards ....½ ounce	Onions, large ....2 ounces	Turnips ....3 ounces

[1] From Countryside & Small Stock Journal, Vol. 73, no. 2 (March/April 1989). Used by permission.

LENGTH OF TIME SEEDS MAINTAIN THEIR VITALITY

Average years	Average years	Average years	Average years
Barley 3	Corn 2	Oats 3	Rape 5
Bean 3	Cucumber, common 6	Onion 2	Rye 2
Beet 6	Eggplant.....................6	Orchard grass 2	Salsify 2
Buckwheat 2	Flax 2	Parsnip........................2	Soybean 2
Cabbage 5	Hop 2	Peanut 1	Squash 6
Carrot 4	Lettuce, common 5	Peas 3	Timothy......................2
Celery 8	Millet 2	Pumpkin 5	Turnip 5
Clover.......................3	Muskmelon 5	Radish..........................5	Watermelon................6
	Mustard 3		Wheat 2

QUANTITY OF SEED PER ACRE

Alfalfa (broadcast) ....20–30 lbs.	Corn....................................6–8 qts.	Rice ....1–3 bu.
Alfalfa (drilled) ....15–20 lbs.	Corn (for silage) ....9–11 qts.	Rye ....3–8 pks.
Barley ....8–10 pks.	Cotton ....1–2 bu.	Sugar beets.....................15–20 lbs.
Beans (field) ....2–6 pks.	Cowpea ....1–1½ bu.	Sweet potato ....1½–4 bu.
Bluegrass (sown alone) ....25 lbs.	Flax ....2–4 pks.	Timothy ....10–20 lbs.
Brome grass	Mangels ....5–8 lbs.	Timothy and clover (mixed)
(sown alone) ....12–20 lbs.	Millet ....1–3 pks.	Timothy ....10–15 lbs.
Buckwheat......................3–5 pks.	Oats ....2–3 bu.	Clover ....4–10 lbs.
Cabbage ....¾–1 lb.	Potato (recommended) ....15–18 bu.	Turnip (broadcast)..............2–4 lbs.
Carrot (for stock) ....4–6 lbs.	Potato ....6–20 bu.	Vetch (hairy),
Clover (alsike alone) ....8–15 lbs.	Pumpkin ....4 lbs.	small grain............ 1 bu. + 1 bu.
Clover (red alone) ....10–18 lbs.	Rape ....2–8 lbs.	Wheat ....6–9 pks.
	Redtop (recleaned) ....12–15 lbs.	

Homemade Fence Tightener

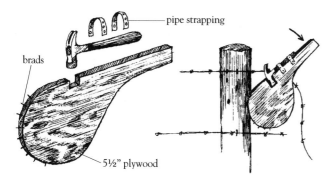

pipe strapping

brads

5½" plywood

VARIOUS CROP SEEDS
Number of Seeds Per Pound —
Standard Weight in Pounds Per Bushel

Crop	Number of seeds per pound	Standard weight per bushel
Alfalfa	220,000	60
Alvee clover	275,000	60
Barley	13,000	48
Bean		
Field	1,000–2,000	60
Lima	400	56
Big trefoil	1,000,000	60
Birdsfoot trefoil	375,000	60
Black medic	300,000	60
		(hulled)
Bluegrass		
Canada	2,500,000	14
Kentucky	2,200,000	14
Brome grass (smooth)	137,000	14
Broom corn	25,000	44–50
Buckwheat		
Common	20,000	48
Tartary	26,000	48
Bur clover		
California (out of bur)	209,000	50
Spotted (in bur)	22,000	8–12
Clover		
Alsike	680,000	60
Crimson	150,000	60
Egyptian (berseem)	210,000	60
Hop	830,000	60
Ladino	860,000	60
Large hop	2,500,000	60

Crop	Number of seeds per pound	Standard weight per bushel
Low hop	860,000	60
Persian	640,000	60
Red	260,000*	60
Strawberry	290,000	60
Subterranean	55,000	60
White	700,000	60
Corn		
Shelled	1,200	56
Ear (husked)	—	70
Pop	3,000	56
Sweet	2,000	50
Cotton	4,000	28–33
Cowpea	2,000–6,000	60
Crested wheatgrass	190,000	20–24
Field peas	3,000–4,000	60
Flax	135,000	56
Hemp	27,000	44
Horsebean	1,000–2,000	47
Johnson grass	130,000	28
Kidney vetch	150,000	60
Lespedeza		
Sericea	335,000	60
Korean	225,000 (unhulled)	40 (unhulled)
Common (Kobe)	190,000 (unhulled)	25 (unhulled)
Common (Tenn. 76)	310,000 (unhulled)	25 (unhulled)
Meadow foxtail	540,000	6–12
Millet		
Foxtail	220,000	50
Japanese	155,000	50
Pearl	85,000	35
Proso	80,000	56
Oatgrass, tall meadow	150,000	11–14
Oats	14,000	32
Orchard grass	590,000	14
Redtop	5,100,000	14
Reed canary grass	550,000	44–48
Rice	15,000	45
Rye	18,000	56
Ryegrass		
Italian	227,000	24
Perennial	330,000	24

Continued on page 106

Crop	Number of seeds per pound	Standard weight per bushel
Sorghum		
Feterita	13,000	56
Hegari	20,000	56
Kafir	20,000	56
Milo	15,000	56
Sorgo	28,000	50
Sorgo (Sumac)	40,000	50
Soybean		
Small-seeded	8,000	60
Medium-seeded	2,000–3,000	60
Large-seeded	1,000	60
Sudan grass	55,000	40
Sweet clover	250,000	60
Timothy	1,230,000*	45
Velvet bean	1,000	60
Vetch		
Common	7,000	60
Hairy	21,000	60
Wheat		
Club	20,000–24,000	60
Common	12,000–20,000	60
Durum	8,000–16,000	60

* George Washington had a penchant for numbers. He felt challenged to calculate the number of seeds in a pound of red clover and a pound of timothy. His conclusions were 71,000 and 298,000, respectively.

Rolling Coulters

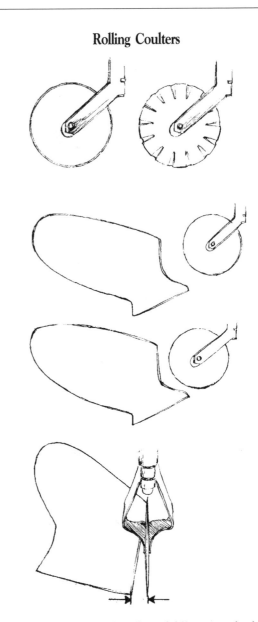

Specific plowing jobs on the farm demand differences in the shape and position of the coulter (a wheel that cuts through sod and soil) as well as differences in the shape and positioning of the moldboard plow.

Pulling Posts

A chain and a notched 2" x 10" board with nailed braces will pull fence posts.

STORAGE REQUIREMENTS AND CHARACTERISTICS OF SEVERAL FRUITS AND VEGETABLES

Kind and Variety or Condition	Recommended storage conditions		Approximate storage life	Average Freezing point, °F	Approximate water content, %	Approximate heat of respiration, Btu/(ton)(24 hr)	
	Temperature °F	Relative humidity, %				At 32°F	At 60°F
Fruits							
Apples							
Jonathan	34–36	85–90	2–4 mo	28.0	84	750	3,000
McIntosh	36–38	85–90	2–5 mo	27.8	84		
Golden Delicious	30–32	85–90	3–6 mo	28.0	84		
Red Delicious	30–32	85–90	3–6 mo	27.7–29.7	84		
Northern Spy	30–32	85–90	4–6 mo	—	84		
Rome Beauty	30–32	85–90	4–7 mo	28.2	84		
Winesap	30–32	85–90	5–8 mo	27.0	84	300	2,300
Apricots	31–32	85–90	1–2 wk	28.9–29.9	85		
Avocados	45 or 55[1]	85–90	4 wk	29.0–31.1	65–82		
Cherries							
Sweet	31–32	85–90	10–14 days	26.2–28.2	80	1,250	
Sour	32	85–90	2–3 days	28.9	84	1,500	12,000
Cranberries	36–40	85–90	1–3 mo	29.7–30.3	87	650	
Figs, fresh	31–32	85–90	10 days	27.1	78		
Grapefruit	—[2]	85–90	1–3 mo	28.3–29.6	89	650	2,600
Grapes							
Vinifera	30–31	85–90	3–6 mo	26.5–27.7	82	400	2,500
American	31–32	85–90	3–4 wk	29.4	82	600	3,500
Lemons	50–58	85–90	1–4 mo	28.5–29.3	89	700	3,650
Limes	48–50	85–90	6–8 wk	27.9–28.6	86		
Oranges							
Florida	30–32	85–90	8–12 wk	26.7–29.7	87	750	4,400
California	35–37	85–90	5–8 wk	29.2	87		
Peaches	31–32	85–90	2–4 wk	29.0–30.1	87	1,100	8,300
Pears							
Bartlett	30–31	90–95	1½–3 mo	28.0	84	750	1,050
Fall and winter	30–31	90–95	2–7 mo[3]	26.5–28.6	83		
Pineapples							
Mature green	50–60	85–90	2–3 wk	29.4–30.0	85		
Ripe	40–50	85–90	2–4 wk				
Plums and prunes....	31–32	85–90	3–4 wk	26.5–29.8	86		
Strawberries	31–32	85–90	7–10 days	30.1–30.5	90	3,250	17,950
Tangerines	31–38	90–95	2–4 wk	29.3–29.7	87		

Continued on page 108

[1] Most varieties at 45°F., some West Indian varieties at 55°F.
[2] Storage temperatures may vary from 32° to 55°F. by area; consult local authority.
[3] Storage period variable for varieties; freezing points about 27°F.

Kind and Variety or Condition	Recommended storage conditions		Approximate storage life	Average Freezing point, °F	Approximate water content, %	Approximate heat of respiration, Btu/(ton)(24 hr)	
	Temperature °F	Relative humidity, %				At 32°F	At 60°F
Vegetables							
Artichokes, Globe	32	90–95	30 days	29.6	84		
Asparagus, green ...	32	85–90	3–4 wk	30.3–30.5	93		
Beans							
Green or snap	45–50	85–90	8–10 days	30.0–30.2	89	5,600	32,100
Lima	32–40	85–90	10–20 days	30.9	67[4]	2,350	22,000
Beets							
Topped	32	90–95	1–3 mo	28.6–30.2	88	2,650	7,250
Bunched	32	90–95	10–14 days	29.5–30.2			
Broccoli, Italian	32	90–95	7–10 days	29.9–30.9	90	7,450	58,100
Brussels sprouts	32	90–95	3–4 wk	30.2	85		
Cabbage							
Early	32	90–95	3–6 wk	30.0			
Late	32	90–95	3–4 mo	—	92	1,200	4,000
Carrots							
Topped	32	90–95	4–5 mo	28.4–29.1	88	2,150	8,100
Bunched	32	90–95	10–14 days				
Cauliflower	32	85–90	2–3 wk	30.1–30.4	92		
Celery	31–32	90–95	2–4 mo	30.7–31.1	94	1,600	8,200
Corn, sweet	31–32	85–90	4–8 days	30.7–30.8	74	6,550	38,400
Cucumbers	45–50	85–90	2–3 wk	29.9–31.0	96	1,700	10,450
Eggplants	45–50	85–90	10 days	30.3–30.4	93		
Endive, or escarole	32	90–95	2–3 wk	30.7–31.7	93		
Garlic, dry	32	70–75	6–8 mo	26.3–30.3	74		
Lettuce	32	90–95	2–3 wk	31.0–31.5	95	11,300	46,000
Melons							
Watermelons	36–40	85–90	2–3 wk	30.3–31.2	92		
Cantaloupes	40–50	85–90	4–14 days	29.6	93	1,300	8,500
Honeydew	45–50	85–90	2–3 wk	29.8–30.0	91		
Onions, dry	32	70–75	6–8 mo	30.0–30.2	82	900	3,000
Parsnips	32	90–95	2–4 mo	29.5–30.1	79		
Peas, green	32	85–90	1–2 wk	29.7–30.5	74	8,200	42,000
Peppers, sweet	45–50	85–90	8–10 days	30.5	92	2,700	8,500
Potatoes							
Early crop	50–70[5]	85–90	—[5]	28.4–30.4			
Late crop	38–50[5]	85–90	5–8 mo[5]	28.7–30.4	78	700	2,000
Pumpkins	50–55	70–75	2–6 mo	29.9	90		
Rutabagas	32	90–95	2–4 mo	29.7	89		
Spinach	32	90–95	10–14 days	31.3	93	4,550	37,500

Continued

[4] Shelled Lima beans.
[5] Variable temperatures and storage life according to areas and uses; consult local authority.

Kind and Variety or Condition	Recommended storage conditions Temperature °F	Relative humidity, %	Approximate storage life	Average Freezing point, °F	Approximate water content, %	Approximate heat of respiration, Btu/(ton)(24 hr) At 32°F	At 60°F
Squashes							
Summer..............	32–40	85–95	10–14 days	29.8–30.9	95		
Winter	50–55	70–75	4–6 mo	29.8–30.0	88		
Sweet potatoes	55–60	85–90	4–6 mo	28.7–29.3	69		
Tomatoes							
Ripe	50	85–90	8–12 days	30.2–30.8	94	1,000	5,650
Mature green	55–70	85–90	2–6 wk	30.6	95	600	6,250
Turnips	32	90–95	4–5 mo	29.8–30.8	91	1,950	5,300

Sources: All information except water contents and freezing points from R.C. Wright, Dean H. Rose, and T.M. Whiteman, *The Commercial Storage of Fruits, Vegetables and Florist and Nursery Stocks*, USDA Agricultural Handbook 66, 1954. Water contents from Charlotte Chatfield and Georgia Adams, *Proximate Composition of American Food Materials*, USDA Circular 549, 1940. Freezing points from T.M. Whiteman, *Freezing Points of Fruits, Vegetables, and Florist Stocks*, USDA Market Research Report 196, 1957. (Range in average freezing points by varieties.)

FERTILIZING ELEMENT BREAKDOWN OF VARIOUS MATERIALS
(by Percentage)

Material	Nitrogen	Phosphoric Acid	Potash
Alfalfa hay	2.45	0.50	2.10
American beech leaves	0.67	0.10	0.65
Ammonium phosphate	12.00	61.00	—
Apple, fruit	0.05	0.02	0.10
Apples, leaves	1.00	0.15	0.35
Apple pomace	0.20	0.02	0.15
Apple skins (ash)	—	3.08	11.74
Balsam fir needles	1.25	0.09	0.12
Barley (grain)	1.75	0.75	0.50
Beet wastes (roots)	0.25	0.10	0.50
Beet wastes	0.40	0.40	3.00
Blood meal	15.00	1.30	0.70
Bonemeal	4.00	21.00	0.20
Brewer's grains (wet)	0.90	0.50	0.05
Brigham tea (ash)	—	—	5.94
Castor-bean pomace	5.50	2.25	1.13
Cattle manure (fresh)	0.29	0.17	0.10
Cherry leaves	0.60	0.11	0.72
Clover hay, alsike	2.05	0.31	1.44
Clover hay, crimson	2.26	0.27	1.86
Clover hay, sweet (white)	2.32	0.29	1.05

Continued on page 110

Material	Nitrogen	Phosphoric Acid	Potash
Clover and timothy hay (mixed)	1.38	0.20	1.58
Coal ash (anthracite)	—	0.125	0.125
Coal ash (bituminous)	—	0.45	0.45
Corncobs (ground, charred)	—	—	2.01
Corncob ash	—	—	50.00
Corn (grain)	1.65	0.65	0.40
Corn (green forage)	0.30	0.13	0.33
Corn stalks	0.75	0.40	0.90
Cottonseed	3.15	1.25	1.15
Cottonseed meal	7.00	2.50	1.50
Cottonseed-hull ashes	—	8.70	23.98
Cotton wastes	1.32	0.45	0.36
Cowpeas, green forage	0.45	0.12	0.45
Cowpeas, seed	3.10	1.00	1.20
Crabgrass, green	0.66	0.19	0.71
Douglas fir bark, ground	0.20	0.09	0.10
Duck manure (fresh)	1.12	1.44	0.49
Eastern hemlock needles	1.05	0.07	0.27
Eggs	2.25	0.40	0.15
Eggshells (burned)	—	0.43	0.29
Eggshells	1.19	0.38	0.14
Feathers	15.30	—	—
Field bean (seed)	4.00	1.20	1.30
Field bean (shells)	1.70	0.30	0.14
Fish scrap (fresh)	6.50	3.75	—
Garbage tankage	1.50	0.75	0.75
Greasewood ashes	—	—	12.61
Garden beans, beans and pods	0.25	0.08	0.30
Garden pea vines	0.25	0.05	0.70
Gluten feed	4.50	—	—
Grape leaves	0.45	0.10	0.35
Grapes (fruit)	0.15	0.07	0.30
Grapefruit skins (ash)	—	3.58	30.60
Greensand	—	1.50	5.00
Hardwood ashes (unleached)	—	2.50	5.50
Hen manure (fresh)	1.63	1.54	0.85
Hoofmeal and horndust	12.50	1.75	—
Horse manure (fresh)	0.44	0.17	0.35
Incinerator ash	0.24	5.15	2.33
Kentucky bluegrass (green)	0.66	0.19	0.71
Kentucky bluegrass (hay)	1.20	0.40	1.55
Larch bark, ground	0.20	0.23	0.13
Leather (acidulated)	7.50	—	—

Continued

Material	Nitrogen	Phosphoric Acid	Potash
Leather (ground)	11.00	—	—
Lemon culls (California)	0.15	0.06	0.26
Mangels	0.22	0.02	0.18
Milk	0.50	0.30	0.18
Millet hay (common)	1.33	0.16	1.78
Mixed grasses, hay	1.22	0.16	1.36
Mussels	0.90	0.12	0.13
Molasses residue in manufacture of alcohol	0.70	—	5.32
Muriate of potash	—	—	48.00
Oak leaves	0.80	0.35	0.15
Oats, grain	2.00	0.80	0.60
Oat straw	0.58	0.09	1.24
Olive pomace	1.15	0.78	1.26
Olive refuse	1.22	0.78	0.32
Orange culls	0.20	0.13	0.21
Orange skins (ash)	—	2.90	27.00
Pea pods (ash)	—	1.79	9.00
Peach leaves	0.90	0.15	0.60
Peanuts (seeds or kernels)	3.60	0.70	0.45
Peanut shells	0.80	0.15	0.50
Peanut shells (ash)	—	1.23	6.45
Pear leaves	0.70	0.12	0.40
Pig manure (fresh)	0.50	0.30	0.50
Pigweed (rough)	0.60	0.16	—
Pine needles	0.46	0.12	0.03
Potassium nitrate	13.00	—	44.00
Potassium phosphate	—	22.00	34.00
Potatoes (tubers)	0.35	0.15	0.50
Potatoes (leaves and stalks)	0.60	0.15	0.45
Potato skins, raw (ash)	—	5.18	27.50
Prune refuse	0.18	0.07	0.31
Pumpkins, fresh	0.16	0.07	0.26
Pumpkin seeds	0.87	0.50	0.45
Rabbit brush ashes	—	—	13.04
Rabbit manure (fresh)	2.40	1.40	0.60
Ragweed, great	0.76	0.26	—
Raspberry leaves	1.35	0.27	0.63
Red clover, green	0.55	0.13	0.50
Red clover, hay	2.10	0.50	2.00
Red maple leaves	0.52	0.09	0.40
Redtop hay	1.20	0.35	1.00
Residues from raw sugar	1.14	8.33	—
Rock phosphate	—	18.00–32.00	—

Continued on page 112

Material	Nitrogen	Phosphoric Acid	Potash
Rhubarb stems	0.10	0.04	0.35
Rye	1.89	0.32	0.47
Salt-marsh hay	1.10	0.25	0.75
Sawdust, alder	0.15	0.09	0.06
Sawdust, Douglas fir	0.05	0.04	0.04
Sawdust, hemlock	0.05	0.04	0.05
Seaweed (Atlantic City)	1.68	0.75	4.93
Sewage sludge from sewer beds	0.74	0.33	0.24
Shavings, alder	0.15	0.09	0.06
Shavings, cedar	0.04	0.04	0.03
Shavings, Douglas fir	0.05	0.04	0.04
Shavings, ponderosa pine	0.04	0.04	0.03
Sheep manure (fresh)	0.55	0.31	0.15
Sludge	2.00	1.90	0.30
Sludge (activated)	5.00	3.25	0.60
Soot from chimney flues	5.25	1.05	0.35
Soybeans	5.84	0.59	2.05
Soybean hay	2.56	0.29	1.93
String bean strings and stems (ash)	—	4.99	18.03
Sugar maple leaves	0.67	0.11	0.75
Superphosphate	—	14.00–20.00	—
Sunflower seed	2.25	1.25	0.79
Sulfate of potash	—	48.00–50.00	—
Sweet potato skins, boiled (ash)	—	3.29	13.89
Sweet potatoes	0.25	0.10	0.50
Tanbark (ash)	—	0.34	3.80
Tanbark ash (spent)	—	1.75	2.00
Tankage	6.00	5.00	—
Timothy hay	1.25	0.55	1.00
Tobacco leaves	4.00	0.50	6.00
Tobacco stalks	3.70	0.65	4.50
Tobacco stems	2.50	0.90	7.00
Tomatoes, fruit	0.20	0.07	0.35
Tomatoes, leaves	0.35	0.10	0.40
Tomatoes, stalks	0.35	0.10	0.50
Triple-phosphate	—	40.00–48.00	—
Turkey manure (fresh)	2.00	1.40	0.60
Urea	46.00	—	—
Vetch hay	2.80	0.75	2.30
Waste from hares and rabbits	7.00	2.40	0.60
Waste from felt hat factory	3.80	—	0.98
Waste product from paint manufacturer	0.02	39.50	—
Waste silt	9.50	—	—

Continued

Material	Nitrogen	Phosphoric Acid	Potash
Wheat, bran	2.65	2.90	1.60
Wheat, grain	2.00	0.85	0.50
Wheat, straw	0.50	0.15	0.60
White ash leaves	0.63	0.15	0.54
White clover (green)	0.50	0.20	0.30
Whey	0.16	0.05	0.21
White sage (ashes)	—	—	13.77
Wood ashes (leached)	—	1.25	2.00
Wood ashes (unleached)	—	1.50	7.00
Wool waste	5.50	3.00	2.00

LIMESTONE REQUIREMENTS
Approximate amounts of finely ground limestone
needed to raise the pH of a 7-inch layer of soil as indicated

Soil regions and textural classes	Limestone requirements — from pH 3.5 to pH 4.5	from pH 4.5 to pH 5.5	from pH 5.5 to pH 6.5	Soil regions and textural classes	Limestone requirements — from pH 3.5 to pH 4.5	from pH 4.5 to pH 5.5	from pH 5.5 to pH 6.5
Soils of warm-temperate and tropical regions:	Tons per acre			Soils of cool-temperate and temperate regions:	Tons per acre		
Sand and loamy sand	0.3	0.3	0.4	Sand and loamy sand	0.4	0.5	0.6
Sandy loam	—	0.5	0.7	Sandy loam	—	0.8	1.3
Loam	—	0.8	1.0	Loam	—	1.2	1.7
Silt loam	—	1.2	1.4	Silt loam	—	1.5	2.0
Clay loam	—	1.5	2.0	Clay loam	—	1.9	2.3
Muck	2.5 [1]	3.3	3.8	Muck	2.9[1]	3.8	4.3

[1] The suggestions for muck soils are for those essentially free of sand and clay. For those containing much sand or clay the amounts should be reduced to values midway between those given for muck and the corresponding class of mineral soil. If the mineral soils are unusually low in organic matter, the recommendations should be reduced about 25 percent; if unusually high, increased by about 25 percent, or even more.

pH READINGS

	pH		pH
Extremely acid	Below 4.5	Neutral[1]	6.6–7.3
Very strongly acid	4.5–5.0	Mildly alkaline	7.4–7.8
Strongly acid	5.1–5.5	Moderately alkaline	7.9–8.4
Medium acid	5.6–6.0	Strongly alkaline	8.5–9.0
Slightly acid	6.1–6.5	Very strongly alkaline	9.1 and higher

[1] Strict neutrality is pH 7.0, but in field work those soils between pH 6.6 and 7.3 are called *neutral*. In the rare cases where significant, the terms *very slightly acid* and *very mildly alkaline* may be used for soils of pH 6.6 to 6.9 and 7.1 to 7.3, respectively.

BUILDING

MACHINERY STORAGE — SPACE REQUIREMENTS

Item	Requirements, ft.		
	Width	Height	Length
Automotive:			
Car	7	6½	18
Tractor:			
One-plow	5	5	9
Two-plow	7½	5	12
Three-plow	7½	5	12
Truck:			
Pickup	7½	7	20
Stock rack	8	10	26
Grain bed	8	7	26
Binder, grain:			
8-ft., reel off	10	5	16
10-ft. tractor, reel on	12	8	19
Corn:			
One-row	7	7	12
Two-row	9	7	16
Baler	13	5–6	21
Bale sleds	6	—	6
Combine:			
5-6 ft.	9–12	8½	16–20
12 ft.	13	10–12	22
16 ft.	20	14	25
Cultivator, corn:			
One-row	5	4	6
Two-row, tractor	10	—	7½
Four-row, tractor	15	—	8
Rotary hoe, two-row	10	3	6
Digger, potato	5	—	8
Drill, grain:			
8-ft., 14–7	11	5	6
10-ft., 16–7	13	5	6
14-ft., 24–7	4	3	9½
Forage Harvester:			
Tractor-drawn,			
2-row corn head	9½	10	15
Windrow pickup attachment	6	4	6
Forage Blower, in transport position			
Long hopper-type	6	6	15½
Short hopper-type	5½	6	8½
Hammer Mill	4	3	9½

Continued

Item	Requirements, ft.		
	Width	Height	Length
Harrow:			
Spike-tooth	4	—	6
Spring-tooth	3	—	6
Disk, horse	8	—	6
Disk, tractor, 8-ft. tractor-mounted	9	2½	9½
Disk, tractor, 7-ft. tractor-mounted	8	4	6
Loader:			
Hay	8	10	12–15
Manure	3–4	6–9	4–10
Mower:			
Horse, bar up	5	6	8
Tractor, trailer-type 7-ft., bar up	5	8	4–6
Tractor, rear-mounted, bar down	12	2	6
Rotary	14	2½	13
Peanut Combine	7½	11	23½
Picker, corn:			
One-row, pull-type	8	6–7	12
Two-row, pull-type	16	6–7	17
Two-row, mounted	10	8	17
Planters, corn or cotton:			
Two-row (without hitch)	10	—	6
Four-row (without hitch)	15	—	6
Potato	6	—	8
Potato Harvester:			
Two-row	14½	11½	25
Plow:			
Walking	2	—	8
Sulky	5	4	7
Two-bottom, horse	5	4	8
Two-bottom, tractor	5–6	4	9½–11
Three-bottom, tractor	6	4	11–13
One-way disk	9	—	10–14
Rack, hay	8	8	16
Rake:			
Dump, 10- and 12-ft.	12–15	4½	6
Side-delivery	8–11	4½	12
Sweep, tractor	9–13	3–4	9–10
Tedder, 8-fork	9	5	6
Seeder, box-type, 11-ft.	13	4	6
Spreader:			
Manure, horse	6	4½	15
Manure, tractor	6	4½	18

Continued on page 116

Item	Requirements, ft.		
	Width	Height	Length
Lime, 8-ft.	10½	—	4
Sugar Beet Harvester	19½	13½	14½
without elevator	14½	4½	14½
Tiller	12–16	—	12
Thresher, grain separator	8–10	10	23–29
Wagon:			
Box and gear, high wheel	6	5½	14
Gear	6	3	9
Box and gear, rubber tire	6	4½	14

CALCULATED DENSITIES OF GRAIN AND SEEDS BASED ON WEIGHTS AND MEASURES USED IN THE DEPARTMENT OF AGRICULTURE

Grain or Seed	Unit	Approximate net weight, lbs.	Bulk density, lbs. per cu. ft.
Alfalfa	bushel	60	48.0
Barley	bushel	48	38.4
Beans:			
Lima, dry	bushel	56	44.8
Lima, unshelled	bushel	32	25.6
Snap	bushel	30	24.0
Other, dry	bushel	60	48.0
Other, dry	sack	100	48.0
Bluegrass	bushel	14–30	11.2–24.0
Broomcorn	bushel	44–50	35.2–40.0
Buckwheat	bushel	48–52	38.4–41.6
Castor beans	bushel	46	36.8
Clover	bushel	60	48.0
Corn:			
Ear, husked	bushel	70 [1]	28.0
Shelled	bushel	56	44.8
Green sweet	bushel	35	28.0
Cottonseed	bushel	32	25.6
Cowpeas	bushel	60	48.0
Flaxseed	bushel	56	44.8
Grain sorghums	bushel	56 & 50	44.8 & 40.0
Hempseed	bushel	44	35.2
Hickory nuts	bushel	50	40.0
Hungarian millet	bushel	48 & 50	38.4 & 40.0
Kafir	bushel	56 & 50	44.8 & 40.0

Continued

Grain or Seed	Unit	Approximate net weight, lbs.	Bulk density, lbs. per cu. ft.
Kapok	bushel	35–40	28.0–32.0
Lentils	bushel	60	48.0
Millet	bushel	48–50	38.4–40.0
Mustard	bushel	58–60	46.4–48.0
Oats	bushel	32	25.6
Orchard grass	bushel	14	11.2
Peanuts, unshelled:			
Virginia type	bushel	17	13.6
Runners, Southeastern	bushel	21	16.8
Spanish			
Southeastern	bushel	25	19.7
Southwestern	bushel	25	19.8
Perilla	bushel	37–40	29.6–32.0
Popcorn:			
On ear	bushel	70[1]	28.0
Shelled	bushel	56	44.8
Poppy	bushel	46	36.8
Rapeseed	bushel	50 & 60	40.0 & 48.0
Redtop	bushel	50 & 60	40.0 & 48.0
Rice, rough	bushel	45	36.0
Rice, rough	bag	100	36.0
Rice, rough	barrel	162	36.0
Rye	bushel	56	44.8
Sesame	bushel	46	36.8
Sorgo	bushel	50	40.0
Soybeans	bushel	60	48.0
Spelt wheat	bushel	40	32.0
Sudan grass	bushel	40	32.0
Sunflower	bushel	24 & 32	19.2 & 25.6
Timothy	bushel	45	36.0
Velvet beans (hulled)	bushel	60	48.0
Vetch	bushel	60	48.0
Walnuts	bushel	50	40.0
Wheat	bushel	60	48.0

[1]The standard weight of 70 lbs. is usually recognized as being about 2 measured bushels of corn, husked, on the ear, because it requires 70 lbs. to yield 1 bu. or 56 lbs. of shelled corn.

SOURCE: ASAE Data: ASAE D241.1, 1967.

DISTANCE TRAVELED PER ACRE WITH EQUIPMENT OF VARIOUS WIDTHS [1]

Width (inches)	Distance traveled per acre (miles)
7	14⅓
8	12¼
9	11
10	9⁹⁄₁₀
11	9
12	8¼
13	7½
14	7
15	6½
16	6⅙
17	5¾
18	5½
19	5¼
20	4⁹⁄₁₀
21	4⁷⁄₁₀
22	4½
23	4¼
24	4
25	4
26	3⅘
27	3⅗
28	3½
29	3½
30	3⅓
31	3⅕
32	3¹⁄₁₀
33	3

Continued

Width (inches)	Distance traveled per acre (miles)
34	2⁹⁄₁₀
35	2⅘
36	2¾
37	2⅔
38	2⅗
39	2½
40	2½
48	2¹⁄₁₂
60	1⅗
72	1⅖

NORMAL ROOT DEPTHS OF SOME MATURE PLANTS [1]

Crop	Feet	Crop	Feet
Alfalfa	5–10	Lettuce	1–1½
Asparagus	6–10	Mint	3–4
Beans	3–4	Onions	1
Cabbage	2	Peas	3–4
Cantaloupes	4–6	Potatoes (Irish)	3–4
Celery	3	Potatoes (sweet)	4–6
Citrus Orchards	4–6	Radishes	1
Corn (field)	4–5	Spinach	2
Corn (sweet)	3	Squash	3
Cranberries	1–2	Strawberries	3–4
Deciduous Orchards	6–8	Tomatoes	6–10
Grain	4	Turnips	3
Grapes	4–6	Watermelons	6

YEARLY PER PERSON CONSUMPTION OF SOME FRUITS AND VEGETABLES[1]

Food	Ave. per capita consumption per year (lbs.)
Potatoes	49.3
Lettuce	24.9
Apples	16
Tomatoes	11.4
Onions, dry	10.2
Cabbage	8.4
Celery	7.3
Cantaloupe	6.6
Sweet corn	6.6
Carrots	6.1
Cucumbers	4.0
Bell peppers	3.3
Pears	2.4
Bananas	20.8

You can, of course, use the chart to evaluate machines of other sizes. If you trade in your 6-ft. hay mower on one that's 10 ft. wide, you know that the 6-footer (72 inches) travels 1⅖ miles to cover an acre. The 10-footer is the same as two fives (60 inches), or half of 1⅗ miles per acre, or about ⅘ of a mile per acre.

[1]From *Countryside & Small Stock Journal*, Vol. 73, no. 2 (March/April 1989). Used by permission.

[1]From *Countryside & Small Stock Journal*, Vol. 73, no. 2 (March/April 1989). Used by permission.

WATER MEASUREMENT TABLE

Cubic Feet per second (c.f.s.)	Gallons per min. (g.p.m.)	Miner's Inches per 24 hours			Acre Inches per hour	Acre Feet per 24 hours
		N. Calif. etc.	S. Calif.	Colorado		
1.	448.8	40.	50.	38.4	0.992	1.983
0.00223	1.	0.0891	0.1114	0.0856	0.0022	0.00442
1.547	694.4	61.89	77.36	59.44	1.535	3.07
0.025	11.25	1.	1.25	0.960	0.0248	0.0496
0.020	9.	0.8	1.	0.768	0.0198	0.0397
0.026	11.69	1.042	1.302	1.	0.0258	0.0516
1.01	452.42	40.32	50.40	38.71	1.	2.
0.504	226.3	20.17	25.21	19.36	0.5	1.

In the above table the number 1 stands for one unit at the rate of flow shown in the corresponding column heading. Read horizontally in either direction to find the equivalent rate in the appropriate column.

TABLE GIVING MINUTE CUBIC FEET OF WATER 1 INCH WIDE FLOWING OVER WEIR

Inches depth over stake	C	⅛ inch	¼ inch	⅜ inch	½ inch	⅝ inch	¾ inch	⅞ inch
1	.40	.47	.55	.65	.74	.83	.93	1.03
2	1.14	1.24	1.36	1.47	1.59	1.71	1.83	1.96
3	2.09	2.23	2.36	2.50	2.63	2.78	2.92	3.07
4	3.22	3.37	3.52	3.68	3.83	3.99	4.16	4.32
5	4.50	4.67	4.84	5.01	5.18	5.36	5.54	5.72
6	5.90	6.09	6.28	6.47	6.65	6.85	7.05	7.25
7	7.44	7.64	7.84	8.05	8.25	8.45	8.66	8.86
8	9.10	9.31	9.52	9.74	9.96	10.18	10.40	10.62
9	10.86	11.08	11.31	11.54	11.77	12.00	12.23	12.47
10	12.71	12.95	13.19	13.43	13.67	13.93	14.16	14.42
11	14.67	14.92	15.18	15.43	15.67	15.96	16.20	16.46
12	16.73	16.99	17.26	17.52	17.78	18.05	18.32	18.58
13	18.87	19.14	19.42	19.69	19.97	20.24	20.52	20.80
14	21.09	21.37	21.65	21.94	22.22	22.51	22.79	23.08
15	23.38	23.67	23.97	24.26	24.56	24.86	25.16	25.46
16	25.76	26.06	26.36	26.66	26.97	27.27	27.58	27.89
17	28.20	28.51	28.82	29.14	29.45	29.76	30.08	30.39
18	30.70	31.02	31.34	31.66	31.98	32.31	32.63	32.96
19	33.29	33.61	33.94	34.27	34.60	34.94	35.27	35.60
20	35.94	36.27	36.60	36.94	37.28	37.62	37.96	38.31
21	38.65	39.00	39.34	39.69	40.04	40.39	40.73	41.09
22	41.43	41.78	42.13	42.49	42.84	43.20	43.56	43.92
23	44.28	44.64	45.00	45.38	45.71	46.08	46.43	46.81
24	47.18	47.55	47.91	48.28	48.65	49.02	49.39	49.76

TYPICAL WINDMILL CAPACITIES[1]

Smaller "fans" (6' to 10') will operate at speeds yielding 32 to 26 strokes of the pump per minute in winds of 15 to 18 mph. Larger fans (12' to 16') yield 21 to 16 strokes per minute in winds of 18 to 20 mph.

The figures below assume that the vane spring is set for full tension, and the fan is running at full speed.

Cylinder Inches	Capacity per Hour, Gallons		Elevation in Feet to Which Water can be Raised					
			Size of Fan					
	6-ft.	8-ft. to 16-ft.	6-ft.	8-ft.	10-ft.	12-ft.	14-ft.	16-ft.
1¾	105	150	130	185	280	420	600	1,000
1⅞	125	180	120	175	260	390	560	920
2	130	190	95	140	215	320	460	750
2¼	180	260	77	112	170	250	360	590
2½	225	325	65	94	140	210	300	490
2¾	265	385	56	80	120	180	260	425
3	320	470	47	68	100	155	220	360
3¼	—	550	—	—	88	130	185	305
3½	440	640	35	50	76	115	160	265
3¾	—	730	—	—	65	98	143	230
4	570	830	27	39	58	86	125	200
4¼	—	940	—	—	51	76	110	180
4½	725	1,050	21	30	46	68	98	160
4¾	—	1,170	—	—	—	61	88	140
5	900	1,300	17	25	37	55	80	130
5¾	—	1,700	—	—	—	40	60	100
6	—	1,875	—	17	25	38	55	85
7	—	2,250	—	—	19	28	41	65
8	—	3,300	—	—	14	22	31	50

[1]From Aermotor, Inc.

POND EARTH DAM HEIGHTS/TOP WIDTHS [1]

Height of Dam (feet)	Minimum top width (feet)
Under 10	8
10 to 15	10
15 to 20	12
20 to 25	14

[1] From USDA.

POND SIDE SLOPES FOR EARTH DAMS[1]

Fill material	Maximum slope	
	Upstream	Downstream
Clayey sand, silty clay, clayey gravel, sandy clay, silty sand, silty gravel	3:1	2:1
Silty clay, clayey silt	3:1	3:1

[1]From USDA.

BLACKSMITH'S STEEL TEMPERING CHART
Tempering Temperatures, Color of Scale, Types of Tools

Degrees F.	Color	Tools
200		
375	Very pale yellow	Gauges
430		Light turning
440	Lemon yellow	Lathes, scrapers
460	Straw yellow	Drills, milling cutters
480	Dark straw	Punches, rock drills, shears
500	Brown	Woodworking, reamers, stone mason tools
510	Brown w/red spots	Wood chisels
520	Brown w/purple spots	Sledgehammers
530	Light purple	Axes, adzes, hot sets, augers, blacksmith's hammers
550	Dark purple	Cold chisels
560	Light blue	Screwdrivers, saws
580	Dark blue	Wagon springs

To temper tool, put in the forge until the proper temperature is reached. To check color properly, tempered area should be cleaned of oxidization and carbon blacking with a steel brush or brick shard. More accurate color evaluations can be had if the tool is held in a shadowed area (like a box). Once the proper color is achieved, quench the tool's section to be tempered in water.

WOODLOT TABLES
Volume in cords (128 cubic feet including bark) by diameter and usable height

Measurement of tree at breast height (inches)		Volume when usable height in feet is —							
Diameter	Circumference	8	16	24	32	40	48	56	64
		Cords	Cords	Cords	Cords	Cords	Cords	Cords	Cords
4	13	0.007	0.011	—	—	—	—	—	—
5	16	.011	.019	0.022	—	—	—	—	—
6	19	.017	.028	.040	0.047	—	—	—	—
7	22	.023	.038	.053	.068	0.076	—	—	—
8	25	.031	.050	.068	.087	.106	0.116	—	—
9	28	.040	.065	.088	.109	.130	.153	0.170	—
10	31	.049	.082	.111	.133	.160	.188	.211	—
11	35	.060	.100	.137	.165	.190	.221	.250	0.270
12	38	.070	.121	.165	.198	.225	.260	.300	.330
13	41	.082	.143	.197	.236	.268	.305	.350	.42
14	44	.095	.167	.228	.273	.311	.353	.40	.47
15	47	.107	.193	.262	.318	.364	.41	.46	.52
16	50	.122	.220	.300	.367	.42	.47	.53	.59
17	53	.138	.250	.340	.42	.48	.54	.59	.66
18	57	.155	.282	.382	.47	.55	.60	.65	.73
19	60	.173	.318	.43	.53	.61	.68	.73	.81
20	63	.194	.353	.48	.59	.68	.76	.81	.89
21	66	.217	.395	.54	.66	.76	.84	.90	.98
22	69	.240	.44	.60	.73	.84	.93	1.00	1.07
23	72	.262	.48	.66	.80	.92	1.03	1.10	1.17
24	75	.288	.52	.72	.88	1.00	1.12	1.21	1.28
25	79	.312	.58	.78	.96	1.10	1.23	1.33	1.38
26	82	.340	.62	.84	1.04	1.19	1.33	1.44	1.51
27	85	.363	.67	.91	1.13	1.29	1.45	1.56	1.63
28	88	.388	.72	.97	1.20	1.38	1.55	1.67	1.76
29	91	.41	.76	1.03	1.29	1.49	1.66	1.80	1.90
30	94	.43	.80	1.10	1.37	1.59	1.7	1.93	2.04

WOODLOT TABLES (continued)
Volume of sawtimber in trees, by diameter and merchantable height. International ¼-inch rule.

Measurement of tree at breast height (inches)		Volume when number of usable 16-foot logs is —								
Diameter	Circumference	1	1½	2	2½	3	3½	4	5	6
		Bd.ft.	Bd.ft.	Bd.ft.	Bd.ft.	Bd.ft.	Bd.ft.	Bd.ft.	Bd.ft.	Bd.ft.
10	31	39	51	63	72	80	—	—	—	—
11	35	49	64	80	92	104	—	—	—	—
12	38	59	78	98	112	127	136	146	—	—
13	41	71	96	120	138	156	168	181	—	—
14	44	83	112	141	164	186	201	216	—	—
15	47	98	132	166	194	221	240	260	—	—
16	50	112	151	190	223	256	280	305	—	—
17	53	128	174	219	258	296	325	354	—	—
18	57	144	196	248	292	336	369	402	—	—
19	60	162	222	281	332	382	420	457	—	—
20	63	181	248	314	370	427	470	522	580	—
21	66	201	276	350	414	478	526	575	656	—
22	69	221	304	387	458	528	583	638	732	—
23	72	244	336	428	507	586	646	706	816	—
24	75	266	368	469	556	644	708	773	899	—
25	79	290	402	514	610	706	779	852	992	—
26	82	315	436	558	662	767	849	931	1,086	—
27	85	341	474	606	721	836	925	1,014	1,185	—
28	88	367	510	654	779	904	1,000	1,096	1,284	1,453
29	91	396	551	706	842	977	1,080	1,184	1,394	1,588
30	94	424	591	758	904	1,050	1,161	1,272	1,503	1,723
31	97	454	634	814	973	1,132	1,254	1,376	1,618	1,862
32	101	485	678	870	1,042	1,213	1,346	1,480	1,733	2,001
33	104	518	724	930	1,114	1,298	1,442	1,586	1,858	2,152
34	107	550	770	989	1,186	1,383	1,537	1,691	1,984	2,304
35	110	585	820	1,055	1,266	1,477	1,642	1,806	2,124	2,458
36	113	620	870	1,121	1,346	1,571	1,746	1,922	2,264	2,612
37	116	656	922	1,188	1,430	1,672	1,858	2,044	2,416	2,783
38	119	693	974	1,256	1,514	1,772	1,970	2,167	2,568	2,954
39	123	732	1,031	1,330	1,602	1,874	2,087	2,300	2,714	3,127
40	126	770	1,086	1,403	1,690	1,977	2,204	2,432	2,860	3,300

Data from Mesavage and Girard, *Tables for Estimating Board-Foot Volume of Timber.* (Form class 80.) U.S. Department of Agriculture, Forest Service. 1946.

For exceptionally tall, slender trees add 10 percent.

For exceptionally short, stubby trees deduct 10 percent.

WOODLOT TABLES (continued)

Contents of logs, in board feet rounded off to the nearest ten, by theInternational log rule, using sawcutting ¼-inch kerf [1]

Diameter of log, small end, inside bark (inches)	Volume when length of log in feet is —						
	8	10	12	14	16	18	20
	Bd. ft.	Bd. ft.	Bd. ft.	Bd. ft.	Bd. ft.	Bd. ft.	Bd. ft.
6	10	10	10	20	20	20	30
7	10	20	20	20	30	30	40
8	20	20	30	30	40	50	50
9	20	30	40	40	50	60	70
10	30	40	50	60	60	80	90
11	40	50	60	70	80	90	110
12	40	60	70	80	100	110	130
13	50	70	80	100	120	130	150
14	60	80	100	120	140	160	180
15	70	90	110	140	160	180	200
16	80	110	130	160	180	210	230
17	100	120	150	180	210	240	270
18	110	140	170	200	230	270	300
19	120	160	190	220	260	300	330
20	140	170	210	250	290	330	370
21	150	190	230	280	320	370	410
22	170	210	260	310	350	400	450
23	190	240	280	340	390	440	500
24	200	260	310	370	420	480	540
25	220	280	340	400	460	530	590
26	240	300	370	430	500	570	640
27	260	330	400	470	540	620	690
28	280	360	430	510	580	660	740
29	300	380	460	550	630	710	800
30	330	410	500	590	670	770	860
31	350	440	530	630	720	820	920
32	370	470	570	670	770	870	980
33	400	500	610	710	820	930	1,040
34	420	530	640	760	870	990	1,110
35	450	570	680	810	930	1,050	1,180
36	480	600	730	850	980	1,110	1,240
37	500	640	770	900	1,040	1,180	1,320
38	530	670	810	950	1,100	1,240	1,390
39	560	710	860	1,010	1,160	1,310	1,460
40	590	750	900	1,060	1,220	1,380	1,540

[1] Source: USDA.

APPROXIMATE AMOUNT OF LABOR NEEDED EACH MONTH FOR SELECTED HOME-PRODUCTION ENTERPRISES IN THE EAST CENTRAL STATES

Enterprise	Hours of labor needed during—												
	Jan.	Feb.	Mar.	Apr.	May	June	July	Aug.	Sept.	Oct.	Nov.	Dec.	Total
Garden, 1 acre, well-diversified, with horse or tractor power........	—	—	30	110	110	50	50	50	50	50	—	—	500
Field corn, 10 acres, cut, shocked, husked by hand	—	—	15	20	35	25	8	—	30	65	32	10	240
Hay, 10 acres	—	—	—	—	—	—	10	—	70	40	—	—	120
Milk cow, 1[1]	20	20	20	20	20	20	10	10	25	20	20	20	225
Laying hens, 12	6	6	6	6	6	6	6	6	6	6	6	6	72
Pigs, 2	—	—	—	8	8	8	8	8	8	—	—	—	48
Bees, 1 or 2 colonies	—	—	—	2	1	2	2	8	2	3	—	—	20
Rabbits, 1 buck and 4 does	10	10	10	10	10	10	10	10	10	10	10	10	120
Milk goats, 2	15	14	14	13	12	12	10	10	10	20	20	15	165

[1]Shift in time of freshening would shift monthly hours of labor in various months.

SOURCE: USDA.

Gate Hardware

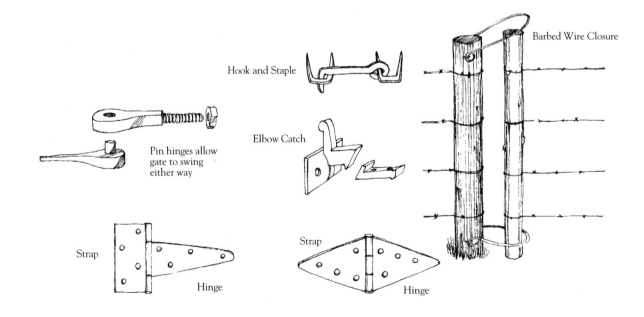

Hook and Staple

Barbed Wire Closure

Pin hinges allow gate to swing either way

Elbow Catch

Strap

Hinge

Strap

Hinge

TABLE OF SIMPLE INTEREST

Principal	Time	4%	5%	6%	7%	8%
$100	1 day	.011	.013	.017	.019	.022
$100	2 days	.022	.028	.033	.039	.044
$100	3 days	.033	.042	.050	.058	.067
$100	4 days	.044	.056	.067	.078	.089
$100	5 days	.056	.069	.083	.097	.111
$100	6 days	.067	.083	.100	.117	.133
$100	7 days	.078	.097	.117	.136	.156
$100	8 days	.089	.111	.133	.156	.178
$100	9 days	.100	.125	.150	.175	.200
$100	10 days	.111	.139	.167	.194	.222
$100	11 days	.122	.153	.183	.214	.244
$100	12 days	.133	.167	.200	.233	.267
$100	13 days	.144	.181	.217	.253	.289
$100	14 days	.156	.194	.233	.272	.311
$100	15 days	.167	.208	.250	.292	.333
$100	16 days	.178	.222	.267	.311	.356
$100	17 days	.189	.236	.283	.331	.378
$100	18 days	.200	.250	.300	.350	.400
$100	19 days	.211	.264	.317	.369	.422
$100	20 days	.222	.278	.333	.389	.444
$100	21 days	.233	.292	.350	.408	.467
$100	22 days	.244	.306	.367	.428	.489
$100	23 days	.256	.319	.383	.447	.511
$100	24 days	.267	.333	.400	.467	.533
$100	25 days	.278	.347	.417	.486	.556
$100	26 days	.289	.361	.433	.506	.578
$100	27 days	.300	.375	.450	.525	.600
$100	28 days	.311	.389	.467	.544	.622
$100	29 days	.322	.403	.483	.564	.644
$100	30 days	.333	.417	.500	.583	.667
$100	1 month	.333	.417	.500	.583	.667
$100	2 months	.667	.833	1.000	1.167	1.333
$100	3 months	1.000	1.250	1.500	1.750	2.000
$100	4 months	1.333	1.667	2.000	2.333	2.667
$100	5 months	1.667	2.083	2.500	2.917	3.333
$100	6 months	2.000	2.500	3.000	3.500	4.000
$100	7 months	2.333	2.917	3.500	4.083	4.667
$100	8 months	2.667	3.333	4.000	4.667	5.333
$100	9 months	3.000	3.750	4.500	5.250	6.000
$100	10 months	3.333	4.167	5.000	5.833	6.667
$100	11 months	3.667	4.583	5.500	6.417	7.333
$100	12 months	4.000	5.000	6.000	7.000	8.000

Borrowing money entails the borrower paying interest (a premium) for the use of that money. Thus, if one borrows $1,000 under the simple interest method (rate percent per annum) for a period of ten years he would pay a premium of $800.

SAMPLE COMMUNITY LAND TRUST LEASE

The following is an abstract from a sample lease used by the Earth Bridge Community Land Trust (Putney, Vermont). Further information can be had by writing to this Land Trust.

1. This lease shall be for a term of ninety years from the date hereof and shall be renewable upon one year's notice in writing to the Lease Committee.

2. The lessee shall hold the land agreeably to the stated purposes of the lessor, as expressed in its Articles of Incorporation and bylaws, as amended from time to time.

3. The lessee may not sublease the premises without the express written consent of the lessor.

4. a. The lessee agrees to pay to the lessor, within thirty days of receipt of notice from the lessor, both the annual property tax or current installment thereof attributable to the leased premises.
   b. The lessee agrees to pay to the lessor annually, an amount determined by the Lease Committee which will be related to the cost to the lessor of administering and maintaining the leased premises. Lessor shall notify lessee in writing of the amount due on or before August 15 and it shall be due on or before September 1.
   c. It is expressly recognized that any approval by the lessor of commercial or industrial use of the premises pursuant to Paragraph 5, may be conditioned on payment of a rental in addition to that described in 4. b., to be agreed to by lessor and lessee.

5. No building construction or other alteration in the land or the use thereof, including timber cutting and significant change in agricultural usage, may be undertaken by the lessee without the express written consent of the Land Use and Building Committee of the lessor. Reasonable conditions may be made a part of that consent.

6. The lessee agrees to consult with and to consider the advice of the designated agricultural expert

of the lessor regarding all future use of the leased premises as a resource.

7. The lessee agrees to permit reasonable inspection of the leased lands by representa-tives designated by the Board of Directors, Land Use and Building Committee or Lease Committee of the lessor.

8. The lessor agrees that this lease may be terminated by the lessee or his/her heirs or assigns at any time provided that written notice of said termination is delivered to the authorized agent of the Lease Committee not less than three months prior to the date of said termination, and provided that no termination shall be effective until a successor lease has been signed. Lessor shall exercise good faith and due diligence in re-letting the premises.

9. Any new structures placed upon the leased land by the lessee shall remain the property of the lessee.

When lining up boundary fence lines from a survey pin at the corner, remember to allow for compass declination.

10. The lessee shall maintain liability insurance on the land, for the benefit of the lessor, with coverage satisfactory to the Lease Committee of the lessor, proof of such insurance to be submitted annually to the Committee on or before September 1 of each year.

11. The lessee shall take care to preserve the purity of water resources on the premises and to insure the disposal of sewage in a safe, legal, and inoffensive manner. The lessee shall comply with any requests by the Lease Committee or Land Use and Building Committee with respect to these matters.

12. The lessee agrees to comply with any procedures with respect to maintenance which the Lease Committee may adopt from time to time.

13. The lessee shall pay an equitable share of the cost of maintenance of facilities used in common with other lessees, the amount to be set by the Lease Committee taking into account the costs involved and the relative benefit conferred on each leasehold.

14. a. Fences shall be maintained by the lessees using them. If an adjoining lessee begins to use a common fence he shall pay to the owner thereof an amount equal to ½ of its current value, and thereafter contribute ½ of the cost of maintenance of said common fence.

    b. Domestic animals owned, controlled, or maintained by the lessee shall not be permitted by the lessee to annoy any neighboring lessee or to graze any neighboring leased parcels in derogation of any other leases with this lessor.

15. a. The lessor reserves the right to use, or lease by separate contract the use of, sugar maples on the leased premises. Lessor and its assigns shall have the right to reasonable access to such maples for sugaring and stand improvement.

    b. The lessee may notify the lessor in writing, on or before January 1, of its intention to use all of specified maple trees during the sugaring season next following, in which case the lessor may not lease such trees. Failure by the lessee to use such trees after giving notice, for three successive seasons, shall constitute waiver of the priority granted under this subparagraph.

16. Lessor expressly reserves to itself all mineral rights, and the use of the property as an extractive resource, and the right to exploit such rights and resources in a reasonable manner.

17. The lessor reserves the right to terminate this lease for the following causes:
    A. Abandonment of the leased property by the lessee.
    B. Subleasing of the leased property without the knowledge and consent of the Lease Committee, including but not limited to, its sublease for residential, commercial, agricultural, or resource-productive purposes.
    C. Violation of any provisions or special restrictions of this lease.
    D. Misuse and/or abuse of this land in violation of this agreement.
    E. Willful or repeated acts to undermine the purposes of the lessor corporation as they are stated in the Articles of Association and bylaws.

SAMPLE ARTICLES OF INCORPORATION FOR A COMMUNITY LAND TRUST

The following are abstracts from the Articles of Incorporation for the Earth Bridge Community Land Trust (Putney, Vermont). Further information can be had by writing to this Land Trust.

Article I: Name

The name of this Corporation shall be Earth Bridge Community Land Trust, Inc.

Article II: Duration

This Corporation shall have perpetual duration and perpetual succession in its corporate name.

Article III: Purposes

This Corporation is organized exclusively for these purposes:

A. To acquire and hold title to property, land, not as public or private property, but in trusteeship, to hold in its stewardship for future generations by keeping said lands from the pressures of the marketplace.

B. To lease said lands on terms that are consistent with sound conservational and ecological principles; terms that reflect this Corporation's interest in the environment and in cooperative planning for associated communities.

C. To collect income from the leased properties.

D. To conserve and maintain abundant organic resources of forest, park, and agricultural lands; to guard against excessive exploitation of income-producing mineral resources and rapid depletion of such resources.

Article IV: Powers and Limitations

This Corporation is empowered to engage solely in activities which serve to further the purposes set forth in Article III above.

This Corporation is not organized for profit and no part of the assets or income of the Corporation shall be distributable to or inure to the benefit of its members, directors, officers, or private persons of any kind, except that the Corporation shall be empowered to pay reasonable compensation for services rendered.

Article VI: Directors

There shall be nine directors elected according to the procedures outlined in the By-Laws of this Corporation.

Article VII: Amendment

These Articles shall not be repealed, amended, or altered in any particular without the unanimous consent of the Directors and the two-thirds majority vote of the membership of the Corporation.

LAND TRUST ORGANIZATIONS

To find out more about conservation easements or other land-protection options for your farm property, the following organizations are good places to start. The national addresses of major trusts and organizations are listed below; for more information on local, statewide, and regional land trust organizations in your area, contact the Land Trust Alliance, below.

A short, useful book on estate planning and tax issues is *Preserving Family Lands* by attorney Stephen J. Small. For more information, contact Mr. Small at Powers & Hall, 100 Franklin Street, Boston, MA 02110-1586.

The American Farmland Trust
1920 N Street, NW, Suite 400
Washington, DC 20036
(202) 659-5170

The American Farmland Trust is a private, nonprofit organization founded in 1980 to protect American farmland. AFT works to stop the loss of productive farmland and to promote farming practices that lead to a healthy environment. Its programs include public education and direct farmland protection projects. Annual membership dues are $15.

The Land Trust Alliance
900 17th Street, NW, Suite 410
Washington, DC 20006
(202) 785-1410

The Land Trust Alliance offers information about, and provides liability insurance for, local land trusts.

The Nature Conservancy
1800 North Kent Street
Arlington, VA 22209
(202) 841-5300

The national headquarters of the Nature Conservancy can provide addresses for the organization's state chapter offices.

PERIODICALS OF INTEREST TO SMALL-SCALE FARMERS

Countryside & Small Stock Journal
N2601 Winter Sports Road
Withee, WI 54498

Countryside & Small Stock Journal is a reader-written journal that "reflects and supports the simple life, and calls its practitioners homesteaders."

Special features touch upon most aspects of husbandry, small-scale agriculture and business, and homestead life. Published bimonthly; subscriptions $18/yr.

The New Farm
222 Main Street
Emmaus, PA 18098

The New Farm "is dedicated to putting people, profit, and biological permanence back into farming by giving farmers the information they need to take charge of their farms and their futures."

Feature articles stress organic and biological alternatives to chemical growing. The magazine is published by the Regenerative Agriculture Association, a program funded by the nonprofit Rodale Institute. Published seven times a year; subscriptions $15/yr.

Organic Farmer
15 Barre Street
Montpelier, VT 05602

Organic Farmer is a newsletter of information on sustainable agriculture. Published quarterly; subscriptions $10/yr.

Sensible Agriculture
Northcutt Communications
P.O. Box 1921
Bothell, WA 98041

Sensible Agriculture is a newsletter dealing with low-input agriculture. Published monthly; subscriptions $39/yr.

Small Farm Technical Newsletter
Whatley Farms, Inc.
P.O. Box 2827
Montgomery, AL 36105-0827

The *Small Farm Technical Newsletter* is dedicated to small farmers and the cash crops that small farms should produce. Information includes all components and production operations from establishment to marketing. Published monthly; subscriptions $12/yr.

Successful Farming
1716 Locust Street
Des Moines, IA 50336

A natural farm and service magazine, published since 1902.

SUGGESTED READING

American Society of Appraisers Staff. *Appraisal of Farmland: Use-Value Assessment Laws and Property Taxation*. Washington, DC: American Society of Appraisers, 1979.

Barnes, Maurice, and Clive Mander. *Farm Building Construction*. Alexandria Bay, New York: Diamond Farm Books, 1986.

Berry, Wendell. *The Gift of Good Land: Further Essays Cultural and Agricultural*. Berkeley, California: North Point Press, 1981.

Bonanno, Alessandro. *Small Farms: Persistence with Legitimation*. Boulder, Colorado: Westview Press, 1987.

Boyd, James S. *Practical Farm Buildings: A Text and Handbook*. 2nd Edition. Danville, Illinois: Interstate Publishers, 1979.

Burch, Monte. *Building and Equipping the Garden and Small Farm Workshop*. Pownal, Vermont: Garden Way Publishing, 1979.

Burch, Monte. *Building Small Barns, Sheds & Shelters*. Pownal, Vermont: Garden Way Publishing, 1983.

Campbell, Stu. *The Home Water Supply: How to Find, Filter, Store, and Conserve It*. Pownal, Vermont: Garden Way Publishing, 1983.

Corley, Hugh. *Organic Small Farming*. San Diego, California: Rateavers, 1975.

Editors of *The New Farm* magazine. *Profitable Farming Now*. Emmaus, Pennsylvania: Rodale Press, 1984.

Faulkner, Edward H. *Plowman's Folly*. Norman, Oklahoma: University of Oklahoma Press, 1943.

Gates, Jane Potter. *Organic Certification*. (USDA Special Reference Brief SRB 90-04.) Beltsville, Maryland: National Agricultural Library, 1990.

Gershuny, Grace, and Joseph Smillie. *The Soul of Soil: A Guide to Ecological Soil Management*. St. Johnsbury, Vermont/Erle, Quebec: Gaia Services, 1986.

Getting Started in Farming on a Small Scale. Brooklyn, New York: Revisionist Press, 1984.

Gregory, Michael, and Margaret Parrish. *Essential Law for Landowners and Farmers*. 2nd Edition. Dobbs Ferry, New York: Sheridan, 1987.

Hess, Oleen. *Small Farms Appropriate Technology*. Beltsville, Maryland: Brandon-Lane Press, 1985.

James, Sidney C., ed. *Midwest Farm Planning Manual*. 4th Edition. Ames, Iowa: Iowa State University Press, 1979.

Jonovic, Donald J., and Wayne D. Messick. *Passing Down the Farm*. Cleveland, Ohio: Jamieson Press, 1987.

Kains, M.G. *Five Acres and Independence: A Practical Guide to the Selection and Management of the Small Farm*. Mineola, New York: Dover Publications, 1973.

Laycock, George. *How to Buy and Enjoy a Small Farm*. New York: McKay, 1978.

Martin, George A., ed. *Farm Equipment and Hand Tools*. Lexington, Massachusetts: Stephen Greene Press, 1980.

McRaven, Charles. *Building with Stone*. Pownal, Vermont: Garden Way Publishing, 1989.

Matson, Tim. *Earth Ponds: The Country Pond Maker's Guide*. Woodstock, Vermont: Countryman Press, 1982.

Midwest Plan Service Personnel. *Farmstead Planning Handbook*. Ames, Iowa: Midwest Plan Service, 1974.

Mulvany, Patrick, comp. *Tools for Agriculture: A Buyer's Guide to Appropriate Equipment*. 3rd Edition.

Croton-on-Hudson, New York: Intermediate Technology Development Group, 1985.

Murray, William G., et al. *Farm Appraisal and Valuation*. 6th Edition. Ames, Iowa: Iowa State University Press, 1983.

MWPS Engineers & Northeast Regional Agricultural Engineering Service. *Small Farms — Livestock Building & Equipment*. Ames, Iowa: Midwest Plan Service, 1984.

Nash, M.J. *Crop Conservation and Storage*. 2nd Edition. Elmsford, New York: Pergamon Press, 1985.

Nearing, Helen, and Scott Nearing. *Continuing the Good Life: Half a Century of Homesteading*. New York: Schocken Books, 1979.

Nearing, Helen, and Scott Nearing. *Living the Good Life: How to Live Simply and Sanely in a Troubled World*. New York: Schocken Books, 1971.

Office for Small-Scale Agriculture Staff. *The Directory for Small-Scale Agriculture*. Washington, DC: Cooperative State Research Service, USDA, 1989.

Preston, Deborah, ed. *Healthy Harvest III: A Directory of Sustainable Agriculture and Horticulture Organizations, 1989-90*. Washington, DC: Potomac Valley Press, 1989.

Roberts, Neal A., and H. James Brown, eds. *Property Tax Preferences for Agricultural Land*. Totowa, New Jersey: Rowman & Allanheld, 1980.

Roy, Ewell P. *Contract Farming and Economic Integration*. Danville, Illinois: Interstate Publishers, 1972.

Sample Application for Farmers Home Administration Self-Help Technical Assistance Grants. Washington, DC: Rural America, 1980.

Schumacher, E.F. *Small Is Beautiful: Economics As If People Mattered*. New York: Harper & Row, 1975.

Shaw, Melvin L. *Cash Crop: Growing Plants the Organic Way*. Wheaton, Illinois: Shaw Publishing, 1986.

Steiner, Frederick. *Ecological Planning for Farmlands Preservation*. Chicago: Planners Press, 1981.

Tompkins, Peter, and Christopher Bird. *Secrets of the Soil*. New York: Harper & Row, 1989.

U.S. Dept. of Agriculture Staff. *Living on a Few Acres*. New York: Plume/NAL, 1979.

Usherwood, N.R., ed. *Transferring Technology for Small-Scale Farming*. Madison, Wisconsin: American Society of Agronomy, 1981.

Vivian, John. *Building Stone Walls*. Pownal, Vermont: Garden Way Publishing, 1978.

Whatley, Booker T., and the Editors of *The New Farm* magazine. *How to Make $100,000 Farming 25 Acres*. Emmaus, Pennsylvania: Regenerative Agriculture Association, 1987.

Index